George Armatage

The sheep

Its varieties and management in health and disease

George Armatage

The sheep
Its varieties and management in health and disease

ISBN/EAN: 9783742842541

Manufactured in Europe, USA, Canada, Australia, Japa

Cover: Foto ©berggeist007 / pixelio.de

Manufactured and distributed by brebook publishing software
(www.brebook.com)

George Armatage

The sheep

Group of St. Kilda Sheep.

THE SHEEP

ITS VARIETIES AND MANAGEMENT IN HEALTH AND DISEASE

REVISED AND ENLARGED

BY

GEORGE ARMATAGE, M.R.C.V.S.

Formerly Lecturer in the Albert and Glasgow Veterinary Colleges

AUTHOR OF "THE HORSE DOCTOR," "THE CATTLE DOCTOR," ETC.

WITH FULL PAGE AND OTHER ILLUSTRATIONS

LONDON

FREDERICK WARNE AND CO.

AND NEW YORK

1894

PREFACE.

The present work is designed to convey to the stock-owner suitable information on the numerous maladies of the sheep. The nature and causes of disease are fully dealt with, and suggestions for prevention are embodied in the text. Indigenous diseases are grouped in accordance with their known characters, as the organs involved, special causes, &c.: and contagious affections are discussed in a separate division. Considerable space is also necessarily devoted to an enumeration of remedies approved in the authorised plan of treatment.

In all respects the reader will find the work is brought up to the stand-point of present-day experience, thus proving a repertory of information on manifold topics, and a safe guide to the sheep-farmer, at home or abroad, in the frequent pressing emergencies with which he has to contend.

Advantage has been taken of an opportunity for revising the contents of this volume. In the chapter devoted to the varieties of sheep, considerable additions have been made, bringing the subject up to date. This part is also greatly enriched by the insertion of illustra tions representing the best types of sheep at the present time. For these the Author is indebted to numerous gentlemen, to whom he desires to express his best thanks. Mr. C. Reid, of Wishaw, supplied the photos, taken by himself, from which Nos. I., II., XV., and XVI. were transferred. The sheep, No. XIV., is inserted by the kind permission of Messrs. Crosby Lockwood & Co., from Professor Wallace's "Farm Live Stock of Great Britain.' Those numbered III., IX., and XII. were courteously supplied by Messrs. Vinton, publishers of the *Live Stock Journal*, and the remainder by the owners or breeders whose names appear in connection with the illustration.'

G. A.

LONDON, 1894

CONTENTS.

CHAPTER I.

THE SHEEP.

PAGE

The Sheep, its class, order, &c.—Origin of the Sheep—The Argali—Characteristics of Sheep—Intelligence—Constitution—Varieties of Sheep—Big-tailed—Big-headed—Wallachian—Iceland Sheep—Value of Sheep—Fat—Flesh—Ewes' milk—Skin—Wool—English breeds of Sheep—Leicester—Wenslydale—Cotswold—Southdown—Hampshire Down—Oxford Down—Shropshire—Welsh and Scotch—Cheviots—Merino Sheep and Wool . . 1

CHAPTER II.

THE SHEEPFOLD.

The breeding flock—Early Lambs—Feeding—Folding—Sheds—Lambing—Instructions to shepherd—Medicine—Puerperal fever—Exercise—Ordinary lambing-time—Breed kept for house Lambs—Management for breeding Lambs in November and December—Feeding the Lambs—A Hampshire farmer's experience—Management of Cheviots and Black-faces on a Lammermoor sheep-farm . . 14

CHAPTER III.

FEBRUARY MANAGEMENT OF SHEEP.

PAGE

Lambing on the Lowland farms—On the Lammermoor farms—Sheep in sheds—Feeding Sheep in yards—Contrasted with Scots—Stall-feeding—Plan of stalls 23

CHAPTER IV.

MARCH, APRIL, AND MAY MANAGEMENT OF THE FLOCK.

Fatting Sheep—How sold to butcher—The Ewe flock—The March lambing—Attention to Ewe—The Lammermoor farm—The April treatment in England and Lammermoor—The flock in May 34

CHAPTER V.

THE FLOCK IN JUNE.

Sheep-washing—Shearing—The fly—Sheep-washes—Recipes for preventing and curing scab—Salving recipe—Arsenic—Australian experience of it—The Ewe flock—Dorset breed—Shearing of Rams—Best Rams—The Ewe flock on the Lammermoor farms 41

CHAPTER VI.

THE FLOCK IN SUMMER.

The flock in summer—Weaning of Lambs—Dipping Sheep—Additional remarks on Sheep-shearing, washing, &c., on the Lammermoor farm—The flock in August—Cheviot Lambs weaned—The flock in September—Purchasing Sheep—Foot-rot—Recipes for cure—The hoose—Curative treatment—Purchase of Ewes and Lambs—The Lammermoor farm 51

CHAPTER VII.

THE SHEEPFOLD IN AUTUMN.

PAGE

The sheepfold in October—Treatment of Tegs—Shearlings fattened for Christmas—The Lammermoor farm—Bathing Sheep—The sheepfold in November—Lammermoor farm—The flock in December—Value of white carrot—Lammermoor memoranda—How to estimate the weight of Sheep—Food and increase 58

CHAPTER VIII.

DISEASES OF SHEEP.

General observations—Value of our domestic animals—Public health dependent upon the health of stock—Study of Sheep diseases frustrated—Opportunities offered—Appeal—Losses, and how they may be averted—Pathology—Symptoms—Morbid pathology—Veterinary medicine—Veterinary surgery—Materia Medica—Fever—Inflammation—Abscess—Serous cyst 65

CHAPTER IX.

Sending for the veterinary surgeon—Imperfect messages—An important query—The conference—"Master has a horse took ill"—The mitigation of suffering—An appeal—Five suggestions—Laconic epistles—"Order is gain" . . . 76

CHAPTER X.

Materia Medica—The actions and uses of medicines, with their forms of combination—Alteratives—Anodynes—Antiseptics or Antiputrescents—Antispasmodics—Astringents—Blisters—Caustics—Clysters or enemas—Cordials—Demulcents—Diaphoretics—Digestives—Diuretics—Electuaries—Embrocations, or liniments—Expectorants—Febrifuges—Fomentations—Lotions—Poultices—Tonics . 83

CHAPTER XI.

PAGE

Blood diseases arising from deranged or inordinate functions —Plethora—Anaemia or "hunger-rot"—Rheumatism—Uræmia—Tubercular consumption—Pining—Apnœa in Sheep and Lambs—Goitre, or "Derbyshire neck"—Rickets, or softening of bone 99

CHAPTER XII.

Diseases of the circulatory system—Anæmic palpitation—Rupture of the heart—Cyanosis—Inflammation of the heart—Foreign bodies in the heart—Pericarditis—Endocarditis—Enlargement of the heart—Dilatation—Fatty degeneration—Displacement of the heart—Embolism . 108

CHAPTER XIII.

Contagious or epizoötic diseases of Sheep—Epizoötic aphtha, or foot-and-mouth disease—small-pox—measles, or rubeola —Means of prevention—Ventilation—Disinfection—Fumigation—Disposal of manure 113

CHAPTER XIV.

Diseases of the digestive organs—Sporadic aphtha, or thrush —Acute indigestion, or hoven—Chronic hoven—Choking —Impaction of the rumen—Foreign bodies in the rumen —Diseases of the second stomach—Dropping the cud—Chronic indigestion — Colic — Diarrhœa — Dysentery—Enteritis, or gastro-enteritis — Peritonitis — Congestion and inflammation of the liver—Jaundice, &c. . . . 124

CHAPTER XV.

Enzoötic diseases caused by animal poisons, non-contagious, but producing a putrid fever in other animals by direct inoculation — Carbuncular erysipelas, or black-quarter—

PAGE

Splenic apoplexy—Gloss Anthrax, or blain—Braxy—Pre-parturient apoplexy—Heaving, or after pains, or parturition fever in ewes 133

CHAPTER XVI.

Enzoötic diseases continued—Malignant catarrh—Arthritis, or joint ill in young lambs—Asthenic hæmaturia, or red or black water—Malignant sore throat—Enzoötic typhoid catarrh, or influenza—Sanguineous dropsy, or red water in the abdomen—Naval ill 142

CHAPTER XVII.

Diseases of the eye—Simple ophthalmia—Iritis and retinitis—Staphyloma—Fungus hæmatodes—Removing the hacks 150

CHAPTER XVIII.

Diseases of the generative organs—Abortion and premature labour—Natural labour—Cleansing—Flooding—Vaginal hæmorrhage—False or unnatural positions of the Lamb, &c.—Inversion of the vagina—Inversion of the womb—Rupture of the womb—Vaginitis—Urethritis—Peritonitis and dropsy of parturition—Garget, or inflammation of the udder—Breaking-down 153

CHAPTER XIX.

Local injuries—Wounds of the abdomen—Breaking-down—Wounds of the bowels—Injuries to the mouth—Wounds of arteries and veins—General wounds—Fractures—Dislocations—Sprains 160

CHAPTER XX.

PAGE

Diseases of the nervous system—Phrenitis, or inflammation of the brain—Apoplexy—Epilepsy—Hydrocephalus, or water in the brain—Paralysis—Hydro-rachitis, or louping-ill—Tetanus, or locked jaw—Rabies 167

CHAPTER XXI.

Parasitic diseases—*Œstrus ovis*, or grub in the nasal sinuses—Hoose or husk, or verminous bronchitis—Gid, or hydatid disease of the brain—Measles—Diseases of the liver due to parasites—Hydatids—Echinococcus parasitism—The fluke disease, or liver rot—Worms in the digestive canal—Scab—Lice, ticks, and maggots 172

CHAPTER XXII.

Poisons—Empirical poisoning—Accidental poisoning—Wilful or malicious poisoning 193

CHAPTER XXIII.

Diseases of the respiratory organs—Simple catarrh, or cold—Sore throat—Bronchitis—Inflammation of the lungs—Abscess in the lungs—Pleurisy—Hydrothorax—Pleuro-pneumonia—Asthma—Enzoötic typhoid catarrh, or influenza 196

CHAPTER XXIV.

Diseases of the skin—Simple inflammation, or erythema—Sore teats—Erysipelas—Simple eczema—Chronic eczema—Impetigo larvalis, or black muzzle—Ecthyma—Weed—Hidebound—Angleberries, or warts—Foot-rot, or *paronychia ovium* 201

CHAPTER XXV.

PAGE

Diseases of the urinary organs—Diabetes—Retention of urine—Incontinence of urine—Simple albuminuria—Hæmaturia, or bloody urine—Sthenic hæmaturia—Inflammation of the kidneys—Inflammation of the bladder—Gravel and calculi, or stone—Protrusion and inversion of the bladder—Weakness of the bladder—Discharges of pus from the bladder 206

Index 213

LIST OF ILLUSTRATIONS

I. GROUP OF ST. KILDA SHEEP	*Frontispiece*	
II. BORDER LEICESTER RAM	*To face p.*	4
III. COTSWOLD SHEARLING RAM, "ROYAL WARWICK" . . .	,,	8
IV. DEVON LONGWOOL TWO-SHEAR RAM .	,,	12
V. DEVON LONGWOOL SHEARLING EWES .	,,	16
VI. LINCOLN CHAMPION RAM	,,	20
VII. LINCOLN SHEARLING LONGWOOL RAM . .	,,	24
VIII. OXFORD DOWN RAM, "HEYTHROP PRINCE II."	,,	28
IX. SHROPSHIRE RAM	,,	32
X. SHROPSHIRE RAM, "TIME WATCH" . .	,,	36
XI. SHROPSHIRE RAM, "ERCALL PRIDE" . .	,,	40
XII. SOUTHDOWN TWO-SHEAR RAM, "SON OF ENTERPRISE"	,,	44
XIII. WENSLEYDALE LONGWOOL EWES	,,	48
XIV. WELSH MOUNTAIN RAM, "BRIHEMIN CYMRU"	,,	52
XV. BLACKFACED (SCOTCH) MOUNTAIN RAM, "LOCHIEL"	,,	56
XVI. BLACKFACED (SCOTCH) MOUNTAIN RAM, "STIRLING"	,,	60

THE SHEEP.

CHAPTER I.

THE SHEEP.

The Sheep, its class, order, &c.—Origin of the Sheep—The Argali—Characteristics of Sheep—Intelligence—Constitution—Varieties of Sheep—Big-tailed—Big-headed—Wallachian—Iceland Sheep—Value of Sheep—Fat—Flesh—Ewe's milk—Skin—Wool—English breeds of Sheep—Leicester—Cotswold—Southdown—Shropshire—Welsh and Scotch—Cheviots—Merino Sheep and wool.

THE Sheep belongs to the fourth class of vertebrate animals; it is of the order Ruminantia—a ruminant, with hollow horns.

The Argali (*Ovis Ammon*) is generally considered as the parent stock of all our varieties of domestic sheep. It is found in great numbers in Kamtschatka, and on the

Head of the Argali or Wild Sheep.

highest mountains of Barbary, of Corsica, and of Greece. It is an agile, active animal, with a very delicate sense of smell, and is captured with difficulty; its flesh is much esteemed. The Argalis prefer mountainous districts, and live in dry and wild places, where they feed on coarse grass and the shoots of young trees. They are very injurious in forests. Their milk is useful as an article of food, and the flesh of their young is eaten.

The principal characteristics of sheep consist in the greater length of their tails, which usually hang down as low as their feet, and also in the bony nature of their horns, which are farther apart at the base, and shaped more spirally than those of the Argali. Further, many breeds of sheep, in both sexes, are entirly destitute of horns.

One thing is certain, that domestic sheep have a very different appearance from their supposed progenitors. The former are possessed neither of the slender or graceful shape nor the nimbleness of pace which is peculiar to the wild breed. The domestic sheep is heavy in its tread and slow in its motions. In them the long and silky hair of the Argali, or Wild Sheep, has almost entirely disappeared; whilst their wool, becoming enormously developed, constitutes a thick fleece. The amount of intelligence they possess is very limited, and their constitution is weak; indeed, they would soon entirely disappear, were it not that man protects them with assiduous and continual care.

In our climate the ewe does not in general produce more than once in a year; but in warmer countries they often bear twice in that period. The length of gestation is five months, and the ewes preserve their milk for seven or eight months after the birth of their young, although the lambs are not allowed to suck for over two or three months. At the age of one year sheep are able to reproduce, and they continue fruitful to the age of ten or twelve years.

Very considerable differences exist in the various varieties of sheep. The Big-tailed Sheep is a breed which is remarkable for the shape of its tail: in them this appendage is expanded to so great an extent with fat, that it often assumes the form of an immense excrescence. This race exists in the temperate parts of Asia, in the south of Russia, in Upper Egypt, and at the Cape of Good Hope. Travellers have stated that in parts of Eastern Africa some of these sheep are harnessed to a kind of small truck, solely for the purpose of supporting the weight of their tails.

There is another race, which is quite as remarkable, known under the name of the Big-headed Sheep. They have no horns, and their necks are supplied with the rudiments of a dewlap, which recalls to mind that of oxen.

The Wallachian sheep is distinguished by its horns pointing straight upwards, and twisting spirally, like those of antelopes.

The Iceland sheep is known to have as many as three, four, and even eight horns.

Sheep are one of the principal sources of agricultural wealth, and furnish, both to commerce and manufacture, products of no inconsiderable importance. Flocks of sheep, from the dung which they leave behind them, are wonderful improvers of the soil. The folding of these animals in a field intended for the cultivation of corn causes beneficial effects which are felt for three consecutive years. Thus their utility in rural economy has long been known. Their wool, for a very considerable period, was considered their most valuable production; but now they supply so vast a quantity of wholesome, agreeable, and very nourishing food, that it is doubtful in which way they most benefit the human family. The fat of sheep, which forms tallow, is likewise one of their most important products; in some breeds it forms a layer from seven to eight inches thick along the ribs and around the loins. Their skin, deprived of the wool, is also applied to numerous purposes. Of this integument are made most of the thin leathers which are used in the manufacture of shoes and gloves. When prepared by other processes, it takes in commerce the names of *chamois, parchment, vellum*, &c. Lastly, milk and cheese are other useful products which are furnished to us by these useful creatures.

Ewe's milk, which is remarkable for its richness, is used in many countries as an article of food, but it is more generally applied to the manufacture of cheese.

The most valuable commodities which are produced by sheep, both in a manufacturing and agricultural point of view, may be summed up as wool and meat. In order to supply these two products in perfection, it is necessary that the animal should present a certain type of conformation.

We shall carefully examine the various varieties of sheep; but before entering upon this subject we will say a few words as to the origin, structure, and qualities of their wool, and the harvest of the fleece.

The sheep's skin produces, in a wild state, two capillary substances: one, stiff and straight, which is called *hair*, and is the most abundant; the other, waving or curled, which is called *wool*, and is the most scanty. In a domesticated state, however, these proportions are reversed; it is the *wool* which is the most plentiful and constitutes the fleece. Under all the efforts of culture the stiff hair tends more and more to decrease. The *fleece* is composed of a collection of *locks* or *slivers*, and the locks of a collection of the *staple*, or hairy fibres.

The *staple* is composed of tubes fitted together, which are only visible in the microscope; their diameter is variable, for which reason it is divided into *extra fine*, *fine*, *middling*, *common*, and *coarse*. Such staple as is equal throughout in diameter, if straight, is much valued; when it is flexuous, the wool is called *wavy;* and when the flexions are very close together, it is pronounced *curly*. This last characteristic appears to belong more particularly to the Merino breed.

The desiderata sought for in wool are *flexibility*, *mellowness*, and *softness;* these properties enable the staple to preserve the qualities which are communicated to it, for then the wool will work or *felt* much more easily, and imparts to the woven fabric the softness and mellowness to the touch which are so much valued. *Elasticity* is also most desirable, for without it wool could not be used in the manufacture of *milled* cloths.

Most of the properties we have just pointed out are due to the greasy matter which penetrates more or less the animal's coat. This lubricating substance is of a very

Border Leicester Ram. The Property of Mr. Samuel Jack, Crighton Mains, Dalkeith.
First and Champion at H. & A. S., Edinburgh, 1893.

complex nature, its composition varying in different breeds. The yolk, for so it is called, is more or less fluid and oily, and is secreted by small glands peculiar to the skin of this race.

When the yolk abounds, it communicates to the wool both softness and pliability; if it is thick and strongly coloured, it imparts to the wool a rough and coarse feel, which necessitates a special process of cleansing or scouring.

Wool is naturally either white, brown, or black. Those of the two last-named colours are less appreciated than the first.

The best wool is found on the sides of the animal's body, from the shoulders to the croup, and underneath as far as the line of the lowest part of the belly.

On the lower part of the belly, where the fleece is less thick (in fact, wanting altogether in some varieties), the locks of wool are felted together, and short, because they are often crushed when the animal lies down.

On the back, the croup, and the top of the thighs, the regularity and uniformity of the locks both diminish, nor does it possess either the mellowness or the pliability of that on the sides. The wool both on the upper and lower parts of the neck is frequently found weak and pendent; that on the head and front of the chest is generally rougher and harsher, as well as being irregular in length and very wavy. The wool on the withers is almost always coarse; that on the ends of the limbs frequently valueless.

Let us now turn to the various breeds of sheep. Of all its varieties and breeds the Leicester probably has hitherto exerted the greatest influence on the merit of sheep stock generally throughout England. It has improved, by its crosses, the long-woolled breeds of the country, more especially the Lincoln and the Cotswold, which retain their names, although they are certainly, to a great extent, altered in character by the admixture of the Leicester blood. It has also induced a useful alteration in the character of the Cheviot breed. In its relation to the Downs, and other short-woolled breeds, it has had most influence as producing a cross-bred animal for feeding, not for breeding from.

The Leicester owes its present improved form to Robert Bakewell, of Dashley, and his successors. He first directed his attention to the improvement of the carcase, which he carried to the extent of rather neglecting the fleece of the sheep.

Such value was ultimately attached to his sheep, that in 1789 he made 1200 guineas by letting three rams, and £2000 by seven others.

The breeds which probably have, most of any, directly been improved by gradual Leicester crossing, are the long-woolled, Lincoln, and Cotswold. The former is the largest sheep in the island.

A three-shear sheep has been known to weigh 96 lbs. a quarter, and fleeces of 16 lbs. and 18 lbs. have been yielded by them.

There are three varieties of Leicester sheep at the present day of which a short account is here given :—

The Border Leicester was originally the outcome of Bakewell's attempt at improvement, by crossing with Cheviot sheep. It is extensively used in the south-east of Scotland, where it ranks as a distinct breed, for crossing with the half-bred Cheviot, as well as black-faced or Highland ewes. Fat tegs of this breed at twelve to fourteen months old, weigh 23 to 25 pounds per quarter.

That which is known as the Improved English Leicester is the smallest of the so-called Leicesters, and possesses the longest record as a distinct improved breed of any, excepting perhaps, the Cotswold. This was successfully accomplished by Bakewell under careful selection and constant inbreeding. In the present day the males are largely used for crossing with the view of improving other breeds. The temperament is mild and the general characteristics favourable to the laying on of fat. The mutton is best at twelve months, when it probably reaches the weight of 20 lbs. per quarter, but the hind quarters are not so large as might be expected.

The Wensleydale is a cross with the Leicester and the

Yorkshire, a large and high-standing animal, exhibiting a shade of blue in the skin of the face and ears, sometimes spreading over the entire body, derived from subsequent crossing with the Scotch black-faced mountain breed. The lambs are extensively used in Scotland for "hogging," and great numbers are annually sold for fatting in the winter, to the graziers of Lincolnshire and East Yorkshire, being known as "Mashams." The mutton is of good quality, and devoid of superfluous fat. This combined with a hardy constitution, though somewhat slow in reaching maturity, places the breed in advance of those in some other districts; but in later years the Border Leicester has been preferred for crossing with Cheviot ewes.

The Cotswolds are generally considered the best of our long-woolled breeds. They possess a large, wide frame; ribs well sprung out; firm, broad, and often overhanging rumps; full hind-quarters; and good thighs; chest full and prominent, but sometimes somewhat defective in depth. The flesh is often rather loose, and not well intermingled with lean. A good Cotswold flock will yield fat mutton, after losing their first coat at fourteen or sixteen months old, 20 lbs. to 24 lbs. a quarter.

The Cotswold ram as well as the Leicester is used to cross the Southdown (the produce is a heavier sheep than the mother, and sells for as much per lb.) but it is best in that case to keep a pure-bred Southdown flock and cross it with the Cotswold ram, and sell the produce not breed from it.

The name given to this breed is said to be derived from the Cotswold Hills, but recent accounts state that as the sheep in olden times were housed in "cotes" or buildings in the "wold," wild or open country, Cotswold became the rendering. Considerable care seems to have been given to the breeding of sheep in that part of Gloucestershire in the days of the Romans. The manufacture of woollen cloth was largely carried on, which, together with sheep, were exported to the Continent. At this day Cotswold sheep also form a large trade with Germany, the United States of America, Australia, and other foreign countries, where they are held in high estimation as pronounced improvers of other breeds.

The race of the Cotswolds has been improved by crossing with the Leicester. It has a hardihood possessed by few others, which enables it to subsist on the produce of stiff, wet land, making fair flesh, and when run over until the age of two years with moderate feeding, the carcase yields as high as 35 lbs. per quarter.

The Southdown is probably the most popular breed in England with the butcher and consumer, and the prices fetched at the annual lettings and sales of such men as Jonas Webb, Rigden, and others, prove that it is still considered one of the most generally useful sorts we have.

The Hampshire Down sheep is said to be the result of a development within the present century from the old Wiltshire horned sheep and the Berkshire Knott, or Nott, by judicious crossing with the improved Southdown. The Wiltshire breed were remarkable for their long, spiral horns, curling close to the head, white faces, and long Roman noses, light forequarters, but wide and heavy in the hindquarters, the fat carcase making 70 to 90 lbs., and the fleece averaging 3 lbs. of wool of medium fineness. The Hampshire Down lambs probably surpass many other breeds for early maturity, attaining in the month of March a dead weight of 12 to 14 lbs. per quarter, and by the month of October, a total weight of 100 lbs. The ram lambs are usually sold in July or August for work as sires, commanding at the auctions of the season from 100 guineas downwards. Their dead weight often reaches 120 lbs.

The Oxford Down originated by a cross with two distinct types of sheep—the longwool and the shortwool. In size and build, as well as weight of fleece, it resembles the first, and in other respects it has approximated the latter, and is classed with the Southdowns and the Hampshire Downs. The breed came into notice in the reign of William IV., and in 1857 it was resolved at a meeting of breeders held in Oxford, to adopt the title of Oxfordshire Downs, since abbreviated to Oxford Downs. It flourishes in many counties of England, as well as in

Cotswold Shearling Ram. The Property of Mr. Robert Garne, Aldsworth, Northleach. First at Warwick, 1892.

Scotland and Wales, adapting itself to the warm and cold climates of countries abroad. The merits of these beautiful creatures, in all the capacities of fleece, meat, early maturity, grazing propensities, and hardness of constitution, are undeniable; from which they have been styled "the glory of the county," and "the general purpose sheep." In Germany the breed has been successfully crossed with the Merino.

Although the higher relative price of the longwools of Leicestershire and the Cotswolds is bringing back the fashion towards these breeds, yet the Southdown and its related breeds still occupy to a great extent the dry arable soil of England. The pure-bred Sussex Down has, like the Leicester upon the long-woolled breeds, exerted an immense influence on all the short-woolled breeds. And thus Hampshire and West Country Downs are sheep of much better quality than they used to be.

Shropshire sheep must be named upon our list as one of the short-woolled class.

The present breed is the result of crossing the Southdown with the Leicester, by which the progeny derived the quality of the first and the size of the second. Subsequent experience has shown that the importation of blood leads to a departure from the true type. The excellent qualities of the race created a demand in almost every county of England, and a few years ago the experiment of using Shropshire rams for crossing the native sheep of Scotland was tried. The result declared that the produce came to early maturity, having developed an excellent covering of flesh on the back, but they do not pay for keeping as tegs until the following spring, being much less in size than the breed they were designed to replace. They likewise proved unsuitable for rearing in the cold regions and hilly pastures, as the lambs during the early days after birth possessed very little wool, and do not resist the cold in the wet and stormy seasons common to the climate.

The Welsh breed and the Cheviots, both fed and reared principally on elevated ground, furnish mutton highly appreciated.

We add in a footnote memoranda of the principal sheep fairs of the country.*

The Merino sheep is, so to speak, a cosmopolitan animal, and may be met with in the most widely divided latitudes, for it has been introduced into Germany, France, England, at the Cape of Good Hope, Australia, Canada, and the United States of America.

The Merino (the word *merino* signifies, in Spanish, *wandering*, and was given probably from the migrations of the Spanish sheep from the mountains to the plains) wool varies in the degrees of fineness, but combines in

* *Blackfaced Breed.*—These are met with at all Scotch fairs, among which the following may be named: Falkirk Trysts, September; Inverness, July; Muir of Ord, April; Slateford, May; Castletown of Braemar, April; Brechin, April and June; Kirriemuir, April and October; and West Linton, June. In England, Stagshawbank and Brough-hill.

Cheviot.—This breed is shown as follows: Lockerbie (lambs), August; Moffat (rams, &c.), September; Langholm (lambs), July, (rams and others) September; Dunse (hoggets in wool and ewes in lamb), March, May (sheep for grazing), July, (lambs and wool), September (draft ewes); Lauder, July; St. Boswell's Green, July; Melrose, August. In England Stagshaw Bank, May; and Brough-hill, September and October.

Leicester Sheep.—Newcastle-on-Tyne, Brough-hill, several Yorkshire fairs, as well as those in Midland counties, as Northampton, Leicester, which are famous. Also Kelso in September.

Southdowns.—In Sussex: Lindfield (lambs), August; Lewes (ewes), September; Horsham (lambs), July; Battle, September. Berkshire: East Ilsley, July, August, and September. Hants: Weyhill, October (principal fair in England for South and Hampshire Downs and Dorset sheep). Wilts: Wilton, September.

Dorset Sheep.—Appleshaw, Hants, October and November; Toller-Down, Dorset, September; and Weyhill, as above.

Cotswolds.—These are shown in the fairs in the counties of Gloucester, Oxford, Berks, and Wilts; at Stow-in-the-Wold, May and October; Burford, April and September; and Marshfield in October.

Romney Marsh.—At Romney (August), and Eastry (October), in Kent. Other longwools, the *Lincoln Sheep* are shown at the fairs of Boston, Gainsborough, Grimsby, Grantham, Partney (August), and other places in Lincolnshire.

The *Exmoor* is obtained at North Devon fairs.

NOTE.—For dates of the above see current Farmers' Almanacs. In many instances the dates are variable, being regulated by some prominent period, feast-day, holiday, &c. For this reason it is considered to be the safest course to omit figures altogether.

the highest degrees, both softness, strength, and elasticity. The fleece covers the whole skin of the animal, down even to its toes, and the tip of the nose is the only part left free. On the other hand, the merino is but indifferent mutton, not only over-burdened with bone, but with a very decided flavour of the wool-grease or yolk.

This breed is probably the most ancient, and in respect of wide distribution, stands in the same position among sheep as the shorthorn among cattle. It is believed to be the progenitor of the sheep of the world through the Spanish Merino, from which have descended the Vermonts, in America; the Negretti of Germany; the Rambouillet of France; the Saxon Merino, and the numerous varieties common to the Australian and Tasmanian Colonies.

Notwithstanding the unfavourable conditions of our northern climate, several breeders of Merinos in this country have continued to preserve their flocks in all their pristine purity. The experience thus gained has established the opinion that Merino wool, grown in this country, is quite equal to that of the best produce of New Zealand, as regards quality, growth and staple. In respect of mutton, great objections are raised against the dark colour, which is somewhat intensified by the freezing process, to which the carcases received from New Zealand, Australia, and the Argentine Republic are subjected. It is urged, however, that the mutton at three years old, when produced on natural food, is quite equal to that of our mountain sheep.

The Merinos are slow in coming to maturity, but crossing with early maturing breeds has the effect of counteracting this defect, an experience largely carried out in connection with the Colonial and South American import trade with this country. In Great Britain the experiment of crossing with the Leicester, Lincoln, Romney Marsh, Shropshire, Cheviot, and Scotch Mountain blackface, has been tried, the last probably being the most successful as referring to climatic conditions, herbage, &c., of the northern pastures,

The following is Professor Simonds' statement of the dentition of the sheep as indicative of age:—

DENTITION OF THE SHEEP.

Table of Early Dentition.			Table of Late Dentition.		
Yrs.	Months.		Yrs.	Months.	
1	0	Central pair of temporary incisors replaced by permanent.	1	4	Two permanent incisors.
1	6	Second pair ,,	2	0	Four ,,
2	3	Third pair ,,	2	9	Six ,,
3	0	Fourth pair ,,	3	6	Eight ,,

Of the relative merits of the different breeds of sheep, as meat manufacturers, strange to say, we have no unexceptionable evidence. It is so difficult to select perfectly representative animals as specimens of the breed to be tried—so necessary to treat each in the manner in which its merits shall most directly appear—so impossible to draw a true comparison of the causes and of effects, when the treatment differs, as it ought to do, with age and breed—that such experiments as have been published cannot be trusted as conveying general truths. Mr. Lawes made an elaborate investigation into the subject, comparing Hampshire and Sussex Downs, Cotswolds, Leicesters, cross-bred wethers, and cross-bred ewes, giving forty of each kind oil-cake, hay, and Swedes during five or six months, and weighing food and increase; and the following, stated generally, are the results arrived at:—

Of the six lots experimented upon, the Cotswolds gave by far the largest average weekly increase per head; indeed, about half as much more than either the Sussex, Leicesters, or cross-bred sheep, and nearly one-fourth more than the Hampshires, which were the second in order of rate of increase per head per week.

Leaving the point of the amounts of food consumed per head, the variations in which, as far as the dry foods are concerned, depended on the varying original weights

DEVON LONGWOOL TWO SHEAR RAM. The Property of MR. E. R. BERRY-TORR, Bideford. First Prize at Torrington, 1892.

of the different lots, and looking only to the amounts consumed per 100 lbs. live weight of animal, or to produce 100 lbs. of increase, it was found that, although the oil-cake and clover chaff were in each case given in proportion to the original weights of the sheep, yet the result was that, taking the average throughout the entire period of the experiment, the Leicesters had less of these dry foods in relation to their average weight than any of the other lots, and more particularly than the Hampshires, Sussex Downs, and Cotswolds. Notwithstanding this, however, the Leicesters also ate less in relation to their average weight of the turnips, which they were allowed *ad libitum*, than any of the other breeds. This less consumption of total food in relation to their weight by the Leicesters might be in their favour, if the result were that they consumed also less for the production of a given amount of increase. But the fact was, that, in relation to the increase they yielded, the Leicesters consumed quite as much food as the cross-breds, and notably more than the Cotswolds. Leicesters, cross-breds, and Cotswolds, however, all gave a larger amount of gross increase for a given amount of food consumed than either the Hampshires or the Sussex sheep. Such were the results of the experiments as they stand on the point of the amount of food required to yield a given amount of increase. But we must not forget that the trials were not all made side by side and in the same season; those with the Hampshires and Sussex Downs being made together in 1850-1, those of the Cotswolds alone in 1851-2, and those with the Leicesters and cross-breds in 1852-3. And although the quality of the respective foods was in all cases as nearly alike as circumstances would allow, yet the actual stocks used were different for the three seasons.

CHAPTER II.

THE SHEEPFOLD.

The breeding flock—Early Lambs—Feeding—Folding—Sheds—Lambing—Instructions to shepherd—Medicine—Puerperal fever—Exercise—Ordinary lambing-time—Breed kept for house-lambs—Management for breeding lambs in November and December—Feeding the Lambs—A Hampshire farmer's experience—Management of Cheviots and Black faces on a Lammermoor sheep farm.

THE BREEDING FLOCK.—Ewes of some breeds of sheep will lamb in January. They should be provided with shelter, food, and attendance. They seldom need turnips till near the end of the year, most farmers having grass sufficient for the ewe flock till they are near lambing, when they should have turnips regularly given them. If the land be not dry, the best method is to draw the turnips, and cart them to a dry pasture, and there bait the sheep on them twice a day, observing well that they eat clean, and make no waste; which is not a bad rule for ascertaining the quantity necessary. In this way the turnip crop goes the farthest. On dry soils, the best way, as the best manuring for the succeeding barley crop, is to eat the crop on the land, hurdling off a certain quantity for the flock; and, as fast as the crop is consumed, whether off the land or cut in troughs, to remove the hurdles farther. This method saves much trouble, and is highly improving to the land; but it should be practised only on lands that are dry, otherwise the sheep poach, and do mischief. The crop, when eaten as it stands, will not go so far as if drawn and laid in a grass-field; for the sheep dung, and stale, and trample on many of the roots after they are begun, which occasions waste; nor is there any loss of manure in carting them, only it is left, in one instance, on the arable field, and, in the other, on the grass one. No improvement can be greater than this of feeding the sheep with turnips. On whatever land they are given, the benefit is always very great.

In wet weather, storms, or snows, the sheep and lambs should be baited on hay, receiving, in moveable racks, a certain quantity every day. And they are the better for a small quantity of hay daily while on turnips, let the weather be good or bad. In some parts, ewes and lambs during January receive bran, malt dust, oats, or oil-cake, in troughs, while they are feeding on turnips. And it has been found capital feeding for them when pulped roots, mixed with straw chaff, have been given them in troughs instead of turnips merely cut or turnips eaten off the land. Take the case of Mr. MacLagan's flock,* described by him as follows:—

"Disapproving of giving them a full supply of turnips, and grudging the expense of feeding them on hay, for which I generally get about £3 per ton, and having always failed in my attempts to make them eat straw, I determined to try the root-pulping system with them. There is generally sufficient grass in my pastures for them till the middle of December. Whenever the grass becomes scanty, I commence to give them pulped turnips and chaff, at the rate of 10 lbs. of turnips to each ewe per day. This is gradually increased to 15 lbs., more than which they seldom get till they are lambed, when they are allowed 20 lbs. and upwards, or, in fact, as much as they can consume. About three weeks before lambing, I mix with the pulped turnips and chaff brewer's grains, bean meal, crushed oats, or some other extra food, to bring the milk upon them; and the same feeding is continued after they have lambed till there is a full bite of grass for them. I also allow them a limited quantity of hay some weeks before they lamb, as my object is to have my lambs fat and ready for market as early as possible."

In severe weather it is right to provide ample shelter, its nature depending on the size, character, and convenience of the farm. The following remarks on this subject are by Mr. Spooner, of Southampton, in a late volume of the *Agricultural Gazette*:—

* Mr. MacLagan of Pumpherstone, Linlithgowshire.—*Journal of the Agricultural Society.*

"For shelter, a large covered shed, fitted with rack and manger for hay and roots, closed on one side, but, with the exception of hurdles, open on the other, will be the most convenient building; and close adjoining there should be a hovel protected on all sides, for the purpose of receiving ewes with weakly lambs, or, in severe weather, to accommodate the lambs as fast as they fall, until they get a little strength. The sheds will, of course, open into the lambing-yard, which should be situated as near as possible to the shepherd's cottage. The yard should face to the south, and should be well bedded with earth, and then littered up with straw, so as to insure cleanliness and afford warmth and comfort to ewes. In the absence of any permanent building, considerable shelter can be afforded in the open field by means of thatched hurdles, to break the force of the prevalent wind (more particularly the east), and also overhead, to keep out the rain or snow. It would also answer well on farms where large flocks are kept to have moveable lambing-houses, such as can be readily taken to pieces and erected again without much trouble.

"Of course, if the flocks are kept in the field, either a hut or a house on wheels will be provided for the shepherd, so that he may not only have shelter for himself, but a fire likewise, with the aid of which he can warm gruel for an exhausted ewe, or prepare any convenient remedy that may be required. The ewes should be visited from time to time during the night, so as to afford assistance when really required, but not to do so officiously: for although in many cases lambs are lost for the want of assistance, yet in others the ewes are sometimes destroyed by unnecessary interference. One rule of importance should be borne in mind, which is, that manual assistance should be rendered to assist, and not to control or oppose, the efforts of Nature. When, therefore, some degree of force is used in removing the lamb, it should be rendered during the labour pains, and it is often needful to wait for their recurrence. The cases most frequently requiring assistance are those where the presentation of the lamb is unfavourable, and where the lamb is dead. The ordi-

DEVON LONGWOOL SHEARLING EWES. The Property of MR. E. R. BERRY-TORR, Bideford. First Prize at Torrington, 1892, and South Molton, 1893.

nary presentation, it is well known, is with the fore-feet first, and the head next, resting on the fore-legs—the parts thus presenting themselves in the form of a wedge. Sometimes the head and at others the legs are bent back, or the fore-feet may be coming together, or the lamb may lie on its back. These are false presentations, and the object should be by means of a small hand and skilful manipulation to turn the lamb, or push back or bring forward the parts that are misplaced. In some cases this cannot be done without destroying the lamb, but it is much preferable to lose the lamb than the ewe. Sometimes the hind parts present first, and then the labour is difficult. With regard to medicine, the following may be given in difficult cases, more particularly when there is much exhaustion:—Opium powdered, 4 drms.; spirit of nitrous ether, 6 oz.; water, 2 oz. Mix. To be well shaken, and a tea-spoonful given as a dose with gruel, in which there has been previously mixed a tea-spoonful of the following powder:—Ginger, 2 oz.; gentian, 2 oz.; cascarilla bark, 2 oz. Mix. When there is much fever the latter may be omitted.*

"Puerperal fever, or heaving after lambing, is a very dangerous disease; indeed, the cases of death preponderate over those of recovery. Inflammatory cases, where there is little exhaustion, may be treated by a copious bleeding at first, and afterwards by sedatives and aperients; but where there is much exhaustion all we can do is to endeavour to rally the efforts of the nervous system by giving the medicine recommended above.

"Though, as tending to abortion, the free use of turnips for heavy ewes is to be avoided before lambing, yet instances of very bad fortune have occurred where turnips had not been given, and other instances of good luck attending where turnips had not been denied. The result of many facts enables us to assign as the cause of warpingand water-bellied lambs the consumption of extremely watery food. With regard to the treatment of abortion, it is a point of much importance to remove the

* See Chapter xviii.

dead lamb as soon as possible; assistance should therefore be given the ewe at each recurrence of the labour pains, and in other respects the treatment should be the same as in parturition.*

"One point is also deserving of much attention, which is, that few things contribute more to the health and well-doing of ewes in lamb than by causing them to take a fair amount of exercise every day."

The feeding flock are drafted off in January, as they are getting fat, and sold with their wool on.

The remarks on the management of lambing ewes so early in the year will appear out of season to the great majority of breeders, whose lambing-time commences in February and extends through March into April; and there are flocks to which they are just as ill-timed from being too late; for when house-lamb is fattened, the lambs must be dropped in November and December, and on their management remarks will be made hereafter. The breed of sheep kept for this purpose is the Horned Dorset, a breed peculiar to the counties of Dorsetshire and Somersetshire. The following details of their management at this season of the year are extracted from a fuller account given in the *Agricultural Gazette*:—

The ewes are brought to take the ram so early as May and June, by feeding them upon trifolium and cut Swedes or mangold placed in troughs, giving them also a change on dry pasture for a few hours during the day; and, if necessary, about half a pint of beans each per day. After the rams are removed from the flock, and the ewes are ascertained to be with lamb, it is best not to keep them too high. A dry pasture, with a change to a fold of tares or similar food, would be most suitable; and they should be managed as a store flock, the object being to keep numbers and to feed the land bare. The travelling consequent upon daily removal for change of food will also prove beneficial; for it is admitted by all flock-masters that the ewes and their offspring will be more healthy when the former have received a moderate amount of exercise during the period of pregnancy.

* See Chapters xv., xvi., and xviii.

Even when the lambing takes place in the months of December or January, it is generally unnecessary to resort to the lambing-yard, because of the comparative mildness of the weather in the counties where this practice of fattening early lamb prevails. On the other hand, it is not advisable to allow the ewes to roam at large during the night-time, as those about to yean are too apt to stray away from the main part of the flock, and their lambs are often lost, or found dead. A shifting fold should be used, being placed on the driest and most sheltered part of the pasture field, and removed on to fresh ground every day. By this means the animals lie on clean land, which, with shelter, will contribute greatly to their health and well-doing, and will, at the same time, enable the shepherd to attend to those ewes which may require assistance; nor can the young lambs, when they fall, escape observation. By this mode of proceeding, the shepherd can at every visit remove all those ewes and lambs which require such care to a place of greater security; for in heavy rain it is necessary to take them to a hovel or covered shed. When the lambs are perfectly strong, and the ewes healthy, it will not be necessary to put them under cover.

If Italian rye-grass has been sown over a certain extent of wheat, then the ewes with their lambs should be placed on the best of this in the wheat stubbles and on the young clovers, taking care to feed the clovers at the day-time, and the wheat stubbles during the night, as the former would receive damage by the stock feeding during the night frosts; and the latter furnishes the best lair and shelter for the young lambs.

In this manner the ewes will give the greatest quantity of milk, and they may be kept upon these grasses until the lamb is a month or five weeks old with immense advantage; for the lambs will be found at the end of that time in the best possible condition. At a month old the ram lambs should be castrated. There are two methods pursued; one called drawing, which is done whilst the lamb is from a week to ten days old; the other called cutting and searing, which may be effected with advantage after the lamb is a month old. The latter plan, though not so commonly followed, is safer; and the lambs, when arrived at maturity, will be found much more fleshy. When the lambs have attained the above-named age, they, as well as the ewes, should be taken from the grass and placed upon root feeding. At this age the lamb begins to require food in addition to its mother's milk; and for the benefit of the ewe it is desirable that the lamb should have it; for although the lamb would go on and improve up to the age of eight or nine weeks old without artificial aid, yet the condition of the mother would be greatly reduced. And it is customary to fat the lamb and the ewe at the same time.

Previous to commencing the feeding of roots, whether they consist of common turnips, Swedes, or carrots, they should be stacked or heaped in readiness for consumption about a week or ten days before being required for use. The advantage of this mode of feeding depends very much upon the roots being cut and placed in troughs, both for the ewes and lambs, which has enabled farmers occupying comparatively cold and heavy land to keep this early stock. Some of our heaviest clay loams, which feed badly in the winter months, produce roots of the best quality; and with the plan of trough-feeding, these soils, under particular management, will produce the best stock, owing to the great feeding value of the roots grown upon them.

The lambs should feed in advance, and separate from the ewes; and therefore a lamb-gate should be provided, with space between the rounds to allow the lambs to pass through freely, without being sufficiently wide to admit the ewes.

The lambs should be fed first, as this will draw them away from the ewes, and otherwise they are apt to contract the habit of feeding with them, which is objectionable, because the ewes receive the coarser food. Feeding should commence as soon as the shepherd can see in the morning; giving hay first, both to lambs and ewes; after which the troughs should be filled with cut roots, taking care to have them cut finest for the lambs, which is done

LINCOLN RAM. Bred by MESSRS. J. RAND, R.R., Kirkham, Biscathorpe; exported by MR. WILLIAM WILSON. Champion at Wangani, New Zealand, 1891.

by passing them twice through the cutter, which plan reduces the food into a state resembling dice, in which state the lambs can readily consume it, and are induced to feed at the earliest period, without loss of time and without waste. As soon as the troughs have been supplied with cut roots, then proceed to give oil-cake and peas, the quantity to be regulated by their wants, always taking care to allow them as much as they will eat. To prevent waste, let the oil-cake be broken fine—about the size of a horse-bean is the best size—otherwise great waste will occur; for the lambs, whilst young, will take large pieces up, and drop them outside the troughs, where it is trodden under-foot and wasted. To induce them to eat cake or peas at first, it is sometimes necessary to mix a small portion of common salt with it. The ewes should next receive their allowance of cake, but without any peas, commencing with a quarter of a pound per day, the half of which should be given at this time, the other half just before the last bait of roots in the evening. After receiving cake for two or three weeks, the quantity may be gradually increased up to a pound per day each, taking care to feed them with only half the full allowance morning and evening; and towards the end of the fattening process half a pint of beans should be given them daily. This renders their flesh more firm, the great objection to ewes fattened while suckling their lambs being, that they are mostly deficient in firmness and quality of meat. Cut roots should be given at times during the day, and the trough quite filled at night.

The lambs whilst young should have hay or hay chaff twice a day; but after they arrive at the age of eight or nine weeks, they should receive hay three times per day; the first bait, as has been stated, the first thing in the morning, the second at noon, and the third about three o'clock in the afternoon. It will not answer much later in the day; for in the short days of winter, after the lambs have drawn away from the ewes, they will lie down for the night, and the portion of hay not consumed will, in case of rain, be distasteful to them, and damaged for further use. The lambs, however, seldom consume all the hay, nor should they be required to do so; for it is better that they should select the best portions of it, the remainder being removed and given to the ewes. In feeding with oil-cake and peas, care should be taken to use covered troughs, and the last bait in the afternoon should not be given later than three o'clock, otherwise a portion may be left in the troughs, which will be damaged in case of rain with change of wind during the night-time. Roots for the lambs should be supplied at short intervals, taking care to have any refuse remaining in the troughs removed every morning; cleanliness in feeding lambs being indispensable. The ewes may receive their oil-cake in open troughs, as they generally eat it immediately they are fed. There is then no time for it to receive damage by rain, &c.; but the troughs should be turned upside down to keep them dry during the time they are not in use.

In illustration of the fitness of these remarks to this season of the year I quote portions of letters written by a Hampshire farmer, on January 6th and January 20th.

"*January 6th.*—Our shepherds are still busy with the lambing season; although our horned ewes finished lambing in the month of December, yet in our flock of Southdown ewes we have at least a moiety which have not lambed; indeed, we consider ourselves warranted in saying that the Southdown ewes of the best and early lambing flocks are at least from ten to fourteen days later to lamb this year."

"*January 20th.*—Our shepherds are now engaged in feeding the ewes and lambs in earnest; the lambing season now drawing to a close, their attention is chiefly turned to fatting the lambs; they are allowed to range over about a quarter of an acre of turnips, and eat off the tops before the ewes; they also get plenty of cut Swedes passed twice through Gardener's machine; they receive also a liberal allowance of white clover hay, and as much linseed cake and cracked peas as they can eat; for we have found, after years of experience, if you would have fat lambs in perfection, they should have plenty of trough food at all times; nor is it less important that the ewes should be well fed, as it is proverbial with us that fat ewes make fat lambs. We have commenced selling our Somerset lambs dropped in October."

How different sheep management is at the two ends of the island may be gathered from the account given to me of a Lammermoor sheep farm consisting of nearly 5,000 acres, at an elevation of 700 to 1,200 feet. The sheep stock amounts to 2,000 ewes and 600 hogs, composed of 800 Cheviot and 1,200 black-faced ewes with the hogs. Most of the black-faced are crossed with Leicester rams; the rest, with Cheviot ewes, are kept pure. The whole produce, except the ewe lambs required to keep up the stock, are sold off. The Cheviot ewes are drafted at four years old, while half of the black-faced are retained until they are five.

It being, of course, a necessary rule that the mouths should not arrive before the food, as spring growth is late on the mountains, the lambing season does not commence till the middle of April, and the tups, in fact, are not taken from among the ewes till the beginning or middle of December, when many Hampshire farms already have lambs six weeks old.

CHAPTER III.

FEBRUARY MANAGEMENT OF SHEEP.

Lambing on the Lowland farms—On the Lammermoor farms—Sheep in sheds—Feeding Sheep in yards—Contrasted with Scots—Stall-feeding—Plan of stalls.

THE lambing season commences on all lowland farms in February, and the constant care and attention of the shepherd are then required.

As this month often proves very severe as regards weather, those ewes which are expected shortly to lamb, should be provided with suitable protection. Sheep and lambs being fatted for market may have full supplies of corn and oil-cake, with hay and straw chaff. Those having foul locks should be cleared of them, and the feet also should be regularly examined for foot-rot. Lame animals do not thrive. On this account the ewes should not be allowed to remain longer than is absolutely needful, and when they are removed to the fields, another item of deep importance is that of ample shelter from wet and stormy winds. This is also the time for "tailing" and castrating the lambs, if the operations were not performed last month, a fine afternoon being needful for the purpose. The ewes thrive well on swedes at this time, and the lambs should have the run of greenfood, as rape, turnips, or thousand-headed kale; corn troughs being placed outside the ewes' pen in order to draw them forward. The ewes benefit considerably by the use of silage.

By way of marking the contrast which obtains between the management of a Highland sheep farm and that proper for February in lowland England, I quote from the Lammermoor sheep farm already mentioned.

"*Lammermoor Sheep Farm, Feb. 5th.*--From the effects of the hard frost on the pastures, the sheep are beginning to lose condition, though they are still good for this time of the year. Towards the end of February the leanest of ewes will be brought in, and get a moderate allowance of turnips daily. With the exception of a few of the weakest, none of the hogs get turnips during the winter, and they are not kept in a separate hirsel, and allowed to graze with the ewes. Two advantages are gained by this. The lambs, when weaned, being only kept from the ewes for about twelve days, when let back to them again generally recognise their own mothers, and continue to follow them during the winter, the ewe scraping the snow and leading to shelter during storms. But the greatest advantage is, that since this plan was adopted there has been comparatively little loss from 'sickness,' an epidemic which, on some farms, in particular seasons, carried off nearly the whole of the young sheep. In some measure to compensate for the want of turnips, all the hogs of both breeds have a flannel jacket, twenty-two inches long

Lincoln Longwoolled Rams. Bred by Messrs. J. R. and R. R. Kirkham, Biscathorpe, Lincoln.

and sixteen broad, sewed along their backs, to keep them warm and dry during cold, wet weather. This plan, as far as we are aware, is only adopted on another farm in this district. Though unsightly at the time, we are convinced that it is of great benefit, doing away with the necessity for smearing with tar, affording more protection to the animal, and greatly improving instead of injuring the fleece. The jackets are put on during the latter end of November, a few days after they are bathed.

"*Feb. 24th.*—At this season hill pastures, being deadened by frosts and bleached with rain, are at their worst, just when the additional demands made on the ewes, which are now heavy with lamb, would require a supply of more nourishing food. Quiet, fresh weather compensates greatly for the want of better food. 'The moss,' or hare's-tail cotton-grass, has been in perfection, and on all grounds where this valuable plant abounds, the sheep are in nearly as good condition as they were three months ago. There is much, however, between the cup and the lip. Yesterday was fine, the ground actually showing symptoms of spring, and our summer visitors, the plovers and curlews, making their appearance, when this day opened with three inches of snow on the ground, and the appearance of severe frosts. Such of the turnips as had escaped the effects of January's frosts have been stored for the use of the Cheviot ewes during March and April. The young sheep were put on turnips on the 22nd, and will continue to receive about one cart-load to 100 sheep, until the lambing has fairly commenced. The old ewes will begin to receive a few about the middle of March. Whenever the ground is sufficiently dry, the shepherds will begin to burn the heather."

Fattening sheep, whether in the field on turnips or other roots, receiving oil-cake or meal at the rate of about 1 lb. apiece, along with chaff of hay and a daily fold of green food, or in sheltered yards, brought in for the more economical consumption of their green food there, will now be making their best progress.

Keep them dry, warm, and clean, if in yards, by daily supplies of fresh litter. The wet and cold weather which generally characterizes this month, makes a great difference between fatting sheep *folded* on turnips and sheep fed on them in yards, in cases where the latter can be kept free from foot-rot. Just the same difference, in fact, as the engineer finds in the cases respectively of a colliery steam-engine, with a boiler exposed to rain and snow, and one such as those found in Cornwall, where every part from which heat can escape is sheltered and covered. The farmer's object should be the same as that of the Cornish engineer. Both must adopt every means to economize the fuel; for this is not with them, as with the engine at the coal-pit, *to be had for nothing*. Turnips and oil-cake are both costly articles—as much so to the farmer as are coals in Cornwall; and the object of those who use them must be to produce the intended effect with as small an expenditure as possible. Sheep and oxen in sheltered spots, and well littered, like steam-engines well "jacketed," waste a less portion of their food or their fuel in keeping up the heat proper to each; and the remaining portion, being much larger, is more effective, by which the force is maintained in the one case, and the fat is laid on in the other. "Warmth is an equivalent for food;" for, supply warmth artificially, and you will not need to supply so much of the food by which it is maintained naturally. Success in feeding depends, no doubt, on a good selection of stock, and on a proper selection of food for them; but it also greatly depends on the attention of the farmer in keeping his stock dry, clean, and *warm*.

Sheep in sheds should this month be making about their best progress. Those intended for the butcher after shearing in May should have received from the 1st Nov., peas, oats, or oil-cake, commencing with half a pint of the first, three-quarters of a pint of the second, or half a lb. of the third; and increasing gradually up to three-quarters of a pint, 1 pint, and 1 lb. respectively each. They will eat with these from 15 up to 25 lbs. of cut Swedes each daily, according to the weight of the animals. It is a fair rule to go by, that an animal when full grown will eat daily of green food a weight equal to one-quarter the weight of its carcase when in fair condition; and it may be assumed that the oil-cake given will reduce the quantity of Swedes required by about 8 lbs. of the latter for every pound of the former. A good crop of Swedes pulled and cut, the sheep being folded on the land, will keep ten sheep for five months per acre; the same crop may be assumed as equal to the keep of thirteen or fourteen under shed. I have had 350 sheep so kept during winter; they ate about 3½ tons of roots daily; and a lad about nineteen years old, with two boys under him, managed the whole. They were placed on two sides of a long yard, which was sheltered on each side, and the space under the shed was divided into pens about 10 ft. by 15; in each of these pens ten sheep were kept. They were littered as often as the straw became wetted, which was about twice a week, and the manure was removed from beneath them about once a month. Their feet were pared once a month; and whenever there appeared the least growth of spongy matter, like that which precedes foot-rot, it was cut, and very dilute nitric acid placed on it. The sheep were fed three times a day, about 8 lbs. of Swedes apiece being given them the first thing in the morning, half a pint of peas about eleven o'clock, 4 lbs. of Swedes at one p.m., and 8 lbs. in the evening.

The practice thus described was faulty, in that the sheep were not littered often enough. It is essential that they be kept dry, in order to avoid the foot-rot; and, if necessary, they should be littered twice a day instead of twice a week.*

The subject of feeding sheep in yards has occupied a good deal of attention; and I extract the following from Mr. Ruston's account of his experience, read before the London Farmers' Club. His farm lies in the fen district; and mangold wurzels are his great resource for yard feeding either of sheep or cattle. He says—

"These fen lands of ours grow a heavy crop of mangolds and a bulky crop of straw, although the quality is inferior. We have the corn standing in the stack-yard,

* See Chapter xxiv.

ready to be threshed, that the straw may be converted, during the winter months, into manure; we have the mangolds also carted into heaps in the neighbourhood of the fold-yard, ready to be consumed; and we have the hay stacked there, too, for the same purpose—at least so much of it as is not required for the work-horses. The question then arises, How can this straw be manufactured into manure, and this accumulation of food be consumed most profitably? Can it be best effected by bullocks or by sheep?

"The more common method of converting our coarse straw into manure, and of consuming our mangolds and inferior hay, has been by purchasing for that purpose some growing bullocks in the autumn, giving them a few pounds of cake or corn per day, in addition to the natural food, and selling them again in the spring either at our home fairs, or at Norwich, or in some other grazing district. If the bullocks have kept healthy and thriven well, they have occasionally left 20*s.* or 30*s.* per head for the natural food consumed; but it has been far more frequently the case that they have only just paid for their artificial food, and the mangolds and hay have had to be charged to the manure account. This mode of management was not very satisfactory. But the case has become even worse during the last few years, since the appearance of the lung disease, and our losses from this cause alone have been fearful.

"I have now tried sheep in yards for five years. Last year the lung disease appearing in a lot of Scots, I had excited my fears lest it should spread and decimate another lot which I had just received from Scotland. I therefore determined at once to send the latter lot away, and sell them again, keeping only those in which the disease had appeared. This drove me to the necessity of purchasing nearly 400 lambs for the purpose of consuming my hay and mangolds, and of manufacturing my straw into manure. I made very close observations, kept a diary of all necessary particulars, valued them into yards, and valued them out to grass, with the dates of going in and out; calculated their cost for arti-

Oxford Down Ram, "Heythrop Prince II." Bred and Exhibited by Mr. Albert Brassey, Heythrop Park, Chipping Norton. First at R.A.S.E., Bath and West, and Oxfordshire Shows, 1893.

ficial food, noticed very narrowly what quantity of straw they made into manure, and also the quality of the manure, as far as appearances enabled me to judge. From close and careful observation last winter, and again this—for I have now between 600 and 700 sheep in yards—I find six lambs will tread down as much straw, and make it into good manure, as a £12 or £14 bullock. I put the sheep into my ordinary fold-yards, and always calculate six sheep to one bullock; so that where I should have ten bullocks I put sixty sheep. During the whole of last winter I don't suppose I had more than a dozen lame sheep whilst they were in the yards; and there have been far less cases in the yards than there were previously to their coming in. I find it very essential to keep a thin layer of dry straw over the yard. In wet days we litter them twice a day, and on fine days once, but we only use a small quantity at a time; this just keeps the heat of the manure from rising to injure their feet, and prevents them also treading on wet straw during the day. When they first come into the yard, and indeed until the end of February, when the days begin to lengthen, we give them a larger quantity of dry food; they pick the bedding straw over, and where practicable have a stack or good heap in the yard to run to; we also cut them chaff, hay, and straw together, and feed them several times a day with it. We give them a few mangolds twice or thrice a day, but not in quantities sufficiently large to make them scour; but as the days lengthen we increase the quantity of roots, and reduce the supply of dry food. I find an acre of mangolds of an average crop will carry twenty-five sheep—*i.e.*, twenty-five lambs—during the weeks they will require to be in the yard, say from the beginning of December to the beginning or middle of April; old sheep would consume more, and twenty per acre should be a fair calculation.

"I will now present some details in connection with my last year's experiments. The 377 lambs wintered in the yards last year were bought during the months of August and September, and were kept entirely at grass-keeping without artificial food, until December, when they were consigned to their winter quarters. A few of them were lost during the winter, but at the end of the winter, when turned out to grass, the following had been the results:—

	£ s. d.	£ s. d.
The whole 377 lambs were valued into the yards at .	618 14 0	
They cost for artificial food .	37 15 1½	
Making a total cost of .		656 9 1½
Further, the 364 put to grass were valued at . . .	891 12 0	
And the 13 casualties realised	9 16 3	
Making a total of . .		901 8 3
Which shows a profit on the whole of		244 19 1½
And this I dispose of as follows—viz., hay, straw, attendance, at 3*s.* per head for 377		56 11 0
15 acres mangold, allowing 25 sheep to the acre, at £12 11*s.* 2½*d.* per acre .		188 8 1½

or, if you take two acres more mangolds, and call the total quantity seventeen acres, it will then give you £11 1*s.* 8*d.* per acre for them, within a fraction; but I regard the former as the more correct calculation."

This was an instance of extraordinary return, and in Mr. Ruston's case it was contrasted with great ill-luck in the cattle feeding with which he had to compare it. This he describes as follows:—

"The lot of Scots to which I have referred as having last year been afflicted with lung disease, were bought on the 12th of March, 1859, and cost just £8 3*s.* 4*d.* per head when they reached my farm. They were put upon a very good (for our country) field of grass, and made considerable progress, and on the 1st of July of the same year, when taking stock with a view to closing and balancing my year's accounts, I valued them at £11 per head. On the 29th of October they were put into yards, and I then valued them at £12 10s. per head. But before that time two of them were seized with lung disease, and had to be killed, and by the 3rd of December the number was reduced to eighteen. Some of the best bullocks fell, and I found the better plan was to dispose of them at once, before they sustained any serious harm. After the 3rd of December no more disease appeared, and I kept the eighteen until the 2nd of February, 1860, when they left me in good health, but went to a bad market, keep being very short last spring, and they brought home, clear of expenses, only £12 11*s.* 6*d.* per head, or just 1*s.* 6*d.* per head more than they were worth on the 29th of the previous October. During the time they were in the yards they consumed chaff—half hay and half straw—and 3 lbs. of the best decorticated cotton-seed cake each per day. They had no mangolds, as they were at a farm where none were grown that year. The cost per head for cake was £1 1*s.* 6*d.* The six that fell with lung disease made £29 6*s.* less than they were valued at when they went into the yards, and the eighteen that did not suffer lost £1 per head on the cake account, besides all the hay, straw, and attendance —rather a costly yard of manure!

"Take, however, a case where 30*s.* per head has been realised for the natural food consumed during the months that cattle have been in the yards, one bullock will return as much profit as two-and-a-half sheep, or a trifle over, and will yet cost as much keeping as six sheep. The figures will stand thus—Profit on one bullock, £1 10*s.*; profit on six sheep, which have consumed the same amount of food, and made the same amount of manure, £3 18*s.* This year, I have, as before stated, 600 sheep in the yards; these are consuming the food and making the manure that 100 bullocks would consume and make. Taking £1 10*s.* as the profit per head on 100 bullocks, and £3 18s. the profit on six sheep—and I think the latter is quite as likely to be realised this year as the former, and indeed more so—what is my position? Why, instead of getting £150 for the food consumed by 100 bullocks, for the very same food consumed by 600 sheep I get £390; which simply puts £240 into my pocket, and emboldens me to argue in favour of sheep as manure manufacturers."

I give Mr. Ruston's experience in detail, as it so

strongly justifies his recommendation to adopt the winter feeding of sheep in yards.

On the quality of the manure thus made it seems plain, as manure is just food *minus* growth, that the kind of animal has nothing to do with the question, which hinges entirely on the quality of food and the kind of growth that is being made out of it. Fatting sheep fed equally well make as good manure as fatting beasts. On the other hand, the cattle are more liable to disease, and it seems that fen-grown food will not fatten them. Sheep, again, if well littered, will not suffer from lameness, and are not liable to any other attack, and they appear to yield a valuable coat of wool over and above as much meat as is made by bullocks. At any rate, in the case of Mr. Ruston's farm, the difference is amazingly in favour of the sheep, and his experience will no doubt induce many copyists of his example.

STALL FEEDING AND SHED FEEDING.—For stall feeding or shed feeding of sheep, again, on a sparred flooring covering a tank for their manure, the outlay need not be excessively great. The following is the plan adopted in a case described in the *Agricultural Gazette* of 1848. The figure opposite represents a section and interior of the arrangement where stall feeding is adopted, each sheep being tied by the neck to its share of the common trough (*b*), which is supplied from a waggon traversing the central railroad (*a*). They stand on an open flooring (*c*), and their dung and urine pass into the tank below. The width of the shed may be from twelve to fourteen feet: each animal will need from eighteen to twenty-four inches in width: the rail above the trough should just touch their shoulders when they stand under it to feed. The roof is of fir poles, the principals being connected by rods of ash, fir, or any other light wood. The uprights are oak posts, refuse of coppice thinnings; the stall rails are hurdle stakes. The tanks are sided with outside slabs, refuse of sawpit; the bottom is the natural bed of earth beaten hard, and half filled with sawdust or burnt earth. The whole of the above is to be found on most farms for the trouble of cutting down. The labour of

Shropshire Shearling Ram. Bred by Mr. W. F. Inge, Thorpe Hall, Tamworth. First and Champion at R.A.S.E., Warwick; and First at Welshpool, 1892.

the sawing of the slabs is hardly to be charged, as it belongs more properly to the planking cut from between the slabs. These materials, if already in existence on an estate, would probably be supplied gratis by the landlord; and for the thatch, heather or straw. The troughs may be of refuse fir slabs, the heavier the better, as the iron

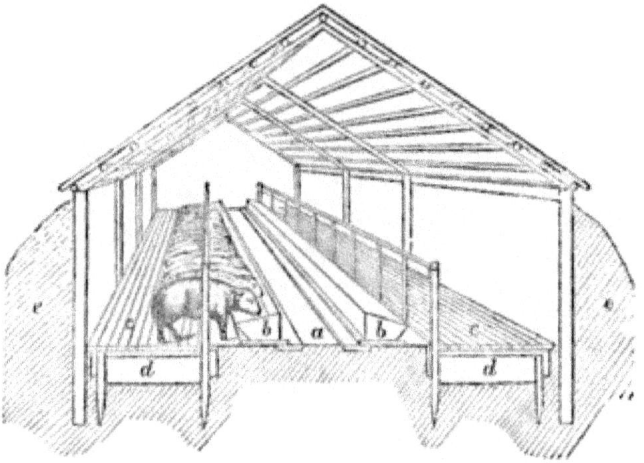

staples of the sheep's collars and chains are driven into them, and clenched inside, and their own weight keeps them steady. The ends of the houses are wattled and thatched, the sides banked up with earth (*e*) taken from the tank, and raised to within six inches of the eaves; the outside slopes should be turfed or seeded.

In cold weather, if additional warmth is desirable, the aperture under the eaves may be closed by long bundles of straw laid lengthwise; and the central opening should also be closed at night by a wattled hurdle. The length of the sheep's chain is from seven to nine inches, long enough to allow the animal to lie down with its head clear of the trough, and not so long as to allow of its putting its fore feet into it. Lastly, for each sparred floor, ten battens of two and three-quarter inches, three-quarters of an inch apart, cover two feet eleven inches, which area is quite sufficient for any sheep. These battens should be of inch larch, with bearings at every three feet, and length to suit the openings of the side uprights.

CHAPTER IV.

MARCH, APRIL, AND MAY MANAGEMENT OF THE FLOCK.

Fatting Sheep—How sold to butcher—The Ewe flock—The March lambing—Attention to Ewe—The Lammermoor farm—The April treatment in England and Lammermoor—The flock in May.

FATTING sheep in fold or yard continue to receive the treatment already described. They may now be shorn as they become fat, and they are thus sold naked to the butcher.

The Ewe flock needs constant care in March, which is in most districts the chief lambing season. We again quote from the *Practical Farmer*:—

"It is requisite to provide suitable lambing paddocks or pens for the lambing season, to which the ewes can be taken every night. In them, and also about the pasture field, or adjoining fields, shelter pens should be constructed, into which a ewe about to lamb, or immediately after she has lambed, should be put, if the weather is unfavourable. These pens are made with straw-wattled hurdles, five hurdles making a double pen, three being set down parallel with each other, and so near that the two other hurdles form the back and front. As the season draws nigh every preparation is madè, and the allowance of food is increased. The ewes being heavy with lamb, require additional support. When their 'time is up,' my ewes are constantly watched for a few days upon the pastures where they have been wintered. As soon as the lambs begin to fall, they are collected into a roomy field, provided as above, in the corner of which, adjoining the shepherd's house, are the lambing paddocks, into which they are driven every dark and unfavourable night. In fine open moonlight nights they are sometimes left out, as it is desirable that the paddocks or lambing pens should be freed from taint, *i.e.*, be sweetened occasionally. Tainted straw or gangrenous droppings are sometimes fatal to ewes if they come in contact with the wounded uterus.* On the large farms upon the Downs and Wolds, and in many other parts of the country, ewes are generally, if not universally, lambed while on turnips. In such cases lambing-yards, or pens of sufficient capacity, should be provided, with a temporary house for the shepherd. It is constant and untiring attention that is required. It is only in special cases that the shepherd's skill and experience are brought into requisition.

"The ewe about to lamb must be examined by the shepherd to ascertain that all is right. This being found correct (*i.e.*, the nose and fore-feet in front, and ready to come forth), she should be left awhile to her own efforts. Should her pains be protracted, the shepherd must lend assistance, or the lamb will be dead ere it is brought forth. The ewe should be gently laid upon her side: the shepherd should then draw one foot forward after the other, and then by a steady pull draw the lamb away. If the lamb is wrongly presented, then comes the difficulty and danger: the shepherd has to push it back, most frequently into the lamb-pouch, there turn it, and by the ewe's pains, and his guidance and help, it is often brought away without much damage. The ewe, in all cases of severe labour, should have a small table-spoonful of laudanum administered, to keep her from paining till the parts become easier, followed in a day or two by a small dose of Epsom salts, and warm cordial drinks or gruel, and suitable food, such as hay, chaff, and sliced turnips or carrots.

"The lamb should be cleared of all adhesion around the nostrils as soon as it is born, and very soon afterwards suckled. The ewe's udder should be cleared of

* See "Heaving Pains," Chapter xv.

wool, to prevent its being drawn in by the lamb. The castration should take place about fourteen days after birth, on a mild, damp morning. As the ewes lamb they should be drafted off to other pastures, and the lambs ought there to have access to a fenced enclosure, where they can enjoy a ration of corn and cake to themselves."

Fatting lambs with their mothers are still folded in the turnip-field, receiving cut turnips, or cut mangolds or carrots in troughs, and they should have also as much peas and oil-cake as they will eat. Some of the forward ones dropped at Christmas-time are now ready for the butcher, and generally make a profitable return for their keep.

Lammermoor Farm.—The following is the report for March from the moorland farm already named :—

"Cheviot ewes are now getting turnips when there is abundance of them. Some are netted close on; others during a part of the day, having a run off during the night; while others have some carted out to their grazing-ground, at the rate of a cart-load to 130 sheep. A few of the leanest of the black-faces are already on turnips, and some more of them will be put on next week. This treatment will be continued till the lambing begins, which will be in about five weeks. Besides the immediate benefit which the sheep derive from the turnips, the pastures will be cleaned and freshened when they return to them, and better enable them to maintain the improved condition they have received from the turnips.

"Spring has hardly yet affected any of the common grasses, except the moss or hare's-tail cotton-grass, the earliest and hardiest of our hill plants, which begins to put out its flower-stems with the first fresh weather in February. It grows most abundantly on spongy, peaty soil. Wherever this valuable plant abounds, sheep not only thrive better, but grow more wool, than those in pastures where it is wanting, even though they get turnips during the spring months. Shepherds are hired at this time for another year. The following are common wages for the district: Thirty ewes, ten ewe hogs, a cow's keep; sixty-four stones (imp.) of oatmeal, six bushels of barley, and 57*s.* in lieu of potatoes. The shepherd manufactures

Shropshire Ram, "Time Watch," 5,624. Bred by and the Property of Mr G. Lewis, Ercall Park, Wellington, Salop. First Prize, R.A.S.E., Doncaster, 1891; First and Champion, Mansell's Memorial Cup (20 guineas). Gold Medal, Market Drayton, 1891; First, Yorkshire Show, Bradford, 1891.

his own peats and hay, the master carting them home; he also furnishes a worker for six weeks as house-rent, or keeps a worker from Whitsunday to Martinmas at 10*d.* per day."

In APRIL the flock receives the same treatment. The lambing season is not yet over, except in early districts. The feeding of fat lamb proceeds, the sheep and lambs being now folded on trefoil and rye, and receiving cut mangolds and peas or oil-cake, or anything else they will eat, *ad libitum.*

In later districts, the experience of the sheep farmer is very different, as may be seen from the following report from the Lammermoor farm :—

Cheviot and black-faced ewes, having been fed on turnips, either at home or elsewhere, have probably returned to their pastures towards the middle of this month. Though there is little spring on the moors, yet having been very lightly stocked for some time, they are clean, so that the ewes do not fall off in condition. Those which are kept on turnips at home, being netted on for seven or eight hours during the day, and allowed to return to the pastures at night, do generally better than those which are in the low country, and constantly confined on the turnips. The first lambs now make their appearance, the black-faced leading the way; indeed, this breed generally come two or three days sooner than either the Cheviots or the Leicesters. The shepherds are now very busy. It is very useful, at this season, to have a few well-sheltered spots, conveniently situated, fenced off some time previous to the lambs making their appearance, into which ewes that are shy to their lambs, or have twins, may be put for a few days. This prevents much confusion and disturbance, as the shepherd knows at once where to find these special objects of his care; and, having been saved for a short time previous, there is generally grass sufficient for them; though when such is not the case, a few turnips, when to be had, will make up the deficiency. The jackets are removed from the hogs about the beginning of April, and the opportunity is taken of branding an initial letter on their horns.

Experience of the jackets confirms the opinion of their utility. Out of 600 ewe hogs (only about seventy of which received ten weeks turnips), the loss during six months has scarcely been one and a half per cent.; while, in respect of condition, they will stand comparison with many in the district which have had four months' turnips. When the jacket is removed, the wool underneath is found to have retained all the yolk it possessed in the end of November, when they were put on, affording a curious contrast to the bleached appearance of those parts which are uncovered. On a hill farm, if the twins are sufficient to fill up the blanks caused by death, it is all that is required. Those ewes having twins, together with some of the worst milkers, are put into enclosed fields, where the grass is rather better, and where hungry lambs can be more easily assisted with cow's milk.

THE FLOCK IN MAY.—The lambing season is now over. The castration of the lamb should be attended to when it is about a fortnight old, and in the later seasons of the hill districts of Scotland much of this operation, especially in the case of the black-faced breed, is accomplished in the month of June. Ewes should be well-treated on the pastures, receiving cut mangolds, and a ration of cake or corn, oats and peas, daily, until there is a full bite of grass, and the lambs are well grown and able to help themselves.

The months of April and May are the great selling-time for fat lamb. We add to our former instructions as to the fatting of lambs the remarks of a correspondent of the *Agricultural Gazette* of many years ago on the state in which it is best to give certain feeding materials. When peas are given to lambs, it is desirable that they should be cracked, not ground into meal; for in this state it is not only objectionable to the lambs, but very wasteful, particularly in damp weather, as much of the finest portion becomes clotted and distasteful, and consequently useless for the purpose intended. It is therefore only necessary that the peas should be broken, and this only up to the time of the lambs being two months old, for after that period they will readily eat them whole.

Good hay may be given to ewes and lambs, either entire or cut into chaff. Fatting sheep (and particularly lambs) should receive their hay in the ordinary state, for they will then have the opportunity to select the best and leave a portion, and afterwards it may be removed and given to the ewes or other stock; whereas, if the hay be given in the state of chaff, the lambs cannot so readily select the best and the clover-leaf portion, nor can it be made so available for the removal of the residue to other stock.

The selection of feeding materials is a matter of the greatest importance; for instance, it is commonly considered that white peas are the best for feeding lambs; but my experience has taught me that the grey or maple varieties are much better than the white. I had an excellent opportunity of proving this a few years ago. Being out of the grey peas, which I usually grew for feeding, I was induced to purchase some of the best white boiling peas. About a week after I commenced feeding with them, my lambs, which had heretofore given good satisfaction, were now complained of by the butcher; nor did they die well and in good condition during the whole time they were eating white peas. But, after awhile, I fed them again with maple peas, when the lambs soon regained their former good quality, and maintained it until the end of the season. It may be considered that the astringent property of the grey and maple pea acts favourably in connection with oil-cake, by conducing to the production of a good proportion of muscle and flesh, which is really essential in making up lambs of the best quality. Beans are not good for feeding lambs, as they contain the astringent property in excess of the peas, and I have known, by their use in feeding, that the flesh has been made so hard as to render it unsaleable as lamb.

There is also a vast difference in oil-cake for feeding purposes. The home-made and some of the Marseilles cake are very good for feeding ewes and sheep stock in general; but the superior sort of American cake is certainly the best for feeding lambs. This cake, when good in quality, always makes a higher price than other sorts of oil-cake; yet it is much cheaper for the feeding of lambs, if we measure its value by results. In fact, I do not hesitate to state that so far as cake and corn are concerned, American barrel cake and maple peas are the perfection of lamb food.

In the growth of grasses intended to produce hay for feeding this kind of stock, it is desirable to select white Dutch clover with trefoil and a small portion of Italian rye-grass mixed for the lambs; and broad clover with trefoil and Italian rye-grass for the ewes. The former being intended as hay for the lambs, should be cut very early indeed; it will then, if well made, contain the greatest amount of nutrition; and this is especially necessary, because young lambs cannot under any circumstances be expected to eat more than a limited quantity. It is therefore requisite that the hay should be of the best quality; for not only will they be induced to eat the greatest bulk of the material, but at the same time the largest possible amount of nutrition will be conveyed into the system. The hay best calculated for feeding the ewes is, without doubt, the same as has been recommended for the lambs; but it often happens that a sufficiency is not grown to feed both with the same sort during the whole season. It is usual to grow the clovers alternately, therefore the supply of either description is somewhat limited, and the broad clover and rye-grass hay is commonly resorted to for the feeding of ewes.

It has been previously recommended to feed the ewes with a certain quantity of oil-cake each per day; it was, however, omitted that they should receive half a pint of beans also, in addition, during the last month of their fattening: this will render the flesh more firm; and they will sell better in the market, inasmuch as the great objection to ewes which have been fattened during the time they suckled their lambs is that they are usually deficient in firmness and quality of meat. One great advantage of high feeding for the ewes is, that the lamb is found to participate in it. The extra quality of the milk induced by feeding upon highly nutritious materials verifies the old saying, "a fat ewe makes a fat lamb."

SHROPSHIRE RAM "ERCALL PRIDE" Bred by and the Property of MR. G. LEWIS, Ercall Park, Wellington, Salop.
First Prize, Bath and West of England Show at Swansea, 1892.

SHEEP-WASHING AND SHEEP-SHEARING, which have been carried on all through the spring as regards fat sheep sold weekly to the butcher, are now carried out upon the great scale, the end of May being the chief time for the operation in the southern part of this island.

CHAPTER V.

THE FLOCK IN JUNE.

Sheep-washing—Shearing—The fly—Sheep-washes—Recipes for preventing and curing scab—Salving recipe—Arsenic—Australian experience of it—The Ewe flock—Dorset breed—Shearing of Rams—Best Rams—The Ewe flock on the Lammermoor farms.

SHEEP-WASHING and sheep-shearing are completed in June. The first object is to provide a convenient place for washing. It is common for men to stand in the water for it, by which they sometimes get bad colds and rheumatic complaints, and must besides be supplied with gin: so disagreeably situated, they hurry over the work, and the wool suffers. A stream or pond offers the requisite opportunity for doing it well, and at the same time comfortably to the men. Rail off a portion of the water for the sheep to walk into by a sloped mouth at one end, and to walk out by another at the other end, with a depth sufficient at one part for them to swim; pave the whole: the breadth need not be more than six or seven feet; at one spot let in on each side of this passage, where the depth is just sufficient for the water to flow over the sheep's back, a cask, either fixed or leaded, for a man to stand in dry. The sheep being in the water between them, they wash in perfection, and pushing them on, they swim through the deep part, and walk out at the other mouth, where is a clean pen, or a very clean dry pasture, to receive them. Of course there is a way to the tubs, and a pen at the first mouth of the water, whence the sheep are turned into it, where they may be soaking a few minutes before being driven to the washers.

The shearing is done after a few days, during which the sheep should be kept in a dry meadow. The yolk rises up into the wool, and both adds to its weight and improves its quality. The cost of clipping will be from 3*s.* to 4*s.* a score, according to the size of the sheep, and in addition to beer.

THE FLY.—Sheep that are kept in enclosures, and especially in a woodland country, should be examined every day, lest they be fly-struck; in twenty-four hours it may be almost past cure.

As a preventive of the fly, train oil is found to be efficacious; but it fouls the wool and makes the sheep disagreeable to touch. An ointment made of butter and the flowers of sulphur seems to be in good repute. When struck, a mercury stone rubbed into the place will destroy the maggots before they have done much mischief.

This may be the proper place and time too (for it is shortly after shearing-time that sheep and lambs are dipped to kill ticks and lice) to refer to the subject of sheep-washes.

We extract the following from the columns of the *Irish Country Gentleman's Newspaper* of July 5th, 1860:—

"By the term 'sheep-wash' the flock-master denotes those compositions which he applies to his flock for the purpose of preventing and remedying scab, the tick, the louse, and the fly. For use as a preventive the solutions are not necessarily made so strong, nor when applied for the destruction of the latter pests, as when used for the cure of that provoking and well-known disease, the scab. The term 'wash,' however, is scarcely definite enough, because for the same purposes Highland flock-masters 'smear' or 'salve' their sheep with compositions in which tar and grease form the principal ingredients; and an unguent composed of mercurial ointment and lard is also successfully used. They are all applied either by rubbing in the compositions, or by immersing the animals in a bath prepared for the purpose. Ointments or salves, of course, must always be rubbed in, and, if applied soon after shearing, the operation does not occupy much time; but if the wool has grown, and it must be closely scored so that the medicine may reach all parts of the skin, it is somewhat tedious. To facilitate the operation, and lest the composition should not be applied with integrity all over the body, the bath has been introduced, at one end of which is placed a horizontal drainer, resting on a shallow wooden trough, which conveys the drainings from the sheep into the bath, so as to prevent waste, the sheep being kept on the drainer for a few minutes. They should not be suffered to go on grass land speedily after being dipped, as disastrous effects have sometimes occurred from animals grazing on pasture where sheep were immediately allowed to roam after leaving the bath. There are some ingenious contrivances for facilitating the operation, by placing the sheep on its back on a kind of cradle, which is suspended by a rope from a sheave, and which can be lowered and raised at option into the bath, so that by merely holding the sheep in suspension over the bath the drainage can be effected to satisfaction.

"There is but little agreement as to the proportions of the constituents of the 'wash,' yet we find all agree that arsenic or mercury must hold a place in the recipe, although many laudable attempts have been made to substitute more innocuous medicines. For the destruction of the tick and louse, however, neither one nor the other need be used; but then the tiny *acari* dwelling within the skin are, in a great measure, saved from the destructive effects of the poison. Non-poisonous remedies have from time to time been trumpeted as discovered; but, as far as we can learn from all the extensive flock-masters we have consulted, the non-poisonous specific for scab is yet a desideratum. Simple cutaneous eruptions yield to sulphur, and most parasitic vermin succumb to tobacco-water; but we have repeatedly proved that neither of these is adequate to curing the scab in sheep.*

"We now give a few of the recipes usually employed for preventing and curing scab, &c.:—Finlay Dun recommends 2 lbs. each of arsenic, pearlash, soft soap, and sulphur, to be suspended in 100 gallons of water.

* See Chapter xxiv., Article "Scab."

This we have tried and found successful, but it discolours the wool.—Youatt's remedy was mercurial ointment, diluted with about five times its bulk of lard. This treatment we have also seen to prove effectual; perhaps its greatest drawback is the difficulty of applying it to all parts of the body, because the least infected spot, if left untouched, will, in the course of time, after the medicine has become *effete*, again spread the contagion over the animal, and subsequently, as a matter of course, over the entire flock.—Hogg's ointment is composed as follows:—

Corrosive sublimate . .	8 oz.
White hellebore, in powder .	12 oz.
Whale or other oil . .	6 gals.
Resin	2 lbs.
Tallow	2 lbs.

These substances should be mixed over the fire, and the sublimate, after being pulverized, should be incorporated with the hellebore and a little oil before being added to the compound. This ointment, though dangerous, is an excellent remedy for scab; but unless the disease actually existed in an aggravated form, we should not give it a preference.—Martin recommends, with caution, ½ lb. of arsenic, mixed in 12 gallons of water. We now detail our own practice in curing scab, which we are in a position to say proves highly effectual:—To 10 gallons of tobacco-water we add—

Corrosive sublimate	3 oz.
White arsenic	3 oz.
Sal ammoniac	3 oz.
Saltpetre	3 oz.
Spirits of turpentine	1 qt.
Starch	1 lb.
Pearlash	1 lb.

The tobacco-water we warm, adding the spirit of turpentine, starch, and pearlash—the starch to keep the minerals in suspension, the pearlash to prevent the yolk from weakening their power. And, subsequently, we add the minerals, after reducing them to a fine powder. This wash we apply, dependent on the season of the year, by hand, from a vessel with a narrow spout, or in

SOUTHDOWN TWO SHEAR RAM, "SON OF ENTERPRISE." Bred by and the Property of MR. H. L. C. BRASSEY, Preston Hall, Aylesford.
First and Champion at Tunbridge Wells; Reserve at R.A.S.E, Chester, 1893.

the bath; in either case the body should be completely wetted, and, where the scab exists, it should be rubbed with a knife, so as to produce a *slight* abrasion. The infected animals of the flock should be first treated, and the sound ones need not have the mixture so strong, which can be regulated by increasing the quantity of tobacco-water. During the time of application, the mixture should be kept stirred. At the same time, all the rubbing places should be washed with the solution, and in about nine days the flock should be gone over again, to see if the scab is lifting off and the itchiness departing, and the skin assuming its ruddy colour—the parts affected by the disease being of a greenish shade. There may be some obstinate cases which will require a second application: the parts should again be slightly abraded, and the solution rubbed in. An ounce of sulphur, given daily, is also found to be of service.

"The practice of salving, so generally adopted in Scotland, might be extended with advantage to many of our hill flocks in this country. The price of the wool, no doubt, is lessened, but the increase in the quantity fully recompenses for this disadvantage, besides the increased health of the animals. Most Highland flock-masters have their own favourite salve; but the following, as recommended by Hogg, is in general use:—

Train or seal oil	4 gals.
Tar	½ gal.
Oil of turpentine	1 pint.

This, mixed, is to be rubbed in after shearing. Dirty butter and tar also are used with effect as a salve. To remedy the depreciation in value of smeared or 'laid' wool, and at the same to secure the advantages which the practice affords, many substitutes have been tried. The most successful we have known is Mr. John Godham's, of Newbigging, who introduced resin instead of tar. His preparation is as follows:—

Butter	18 lbs.
Hogs' lard	18 lbs.
Resin	12 lbs.
Gallipoli oil	1 gal.

"An excellent mixture, to prevent ticks and fly-striking, may be prepared as follows :—To 10 gallons of tobacco-water add 4 oz. of arsenic, ½ lb. of soft soap, and 1 oz. of assafœtida, to be prepared and supplied as the solution we have ourselves recommended.—Powdered white lead, applied to the parts where the maggots are found, is a capital remedy; the maggots, however, should be first removed by a pointed stick. But the scab-water we recommend is equally effectual, if poured on the part.—We have lately examined the sheep-wash compound prepared by the Southdown Company, and have found it to be a concentrated essence of tobacco. This compound is in extensive use in the United States.

"It is worthy of remark that in these climates sheep can bear much more arsenic or mercury than in warmer and drier climates. The same quantities of these medicines which may be applied with impunity here would be attended with the most disastrous results in Australia."

On the use of arsenical solutions for this purpose, it appears, indeed, from Australian experience, that in hot weather they are absorbed and act as poison, while in cold weather they are harmless, and act only in the manner for which they are applied. Mr. Young, of Kirkliston, thus speaks of his Australian experience :—

"There are two solutions which I have been in the habit of using, although not equally extensively, for reasons which will presently appear. One of these is made by dissolving from ½ oz. to 6 drms. of arsenic in a gallon of water; and in a solution of this strength I have dipped as many as 2,000 sheep a day, with, during cold weather, no fatal consequences ensuing. In warm weather, however, and especially if the sheep be in fine condition, the use of this solution is very dangerous; so much, that in one instance, on using, during the summer, a solution containing only 2 drms. of arsenic to the gallon of water, ten per cent. of the sheep (fat wethers) died. And as another instance, I may mention that a neighbour, in December of last year (summer-time in Australia), dipped a flock of picked ewes and lambs in an arsenical solution, the strength of which, however, I do not know, and the result was that 1,300, being exactly one-half of them, died.

"The application of these remarks to a recent case, Black *v.* Elliott, wherein nearly all the plaintiff's sheep died after dipping, is plain. They were in fine condition, and were dipped, during the month of August, in a solution containing ½ oz. of arsenic to the gallon of water. The result was to have been expected." *

THE EWE FLOCK.—The ewes from which it is proposed to breed early lambs for fattening early in the year should have the ram with them. Indeed, for the earliest lambs the ram is put with them in the month of May. The following remarks are from the same pen as described this management in early spring.

The breed of sheep kept for the rearing of early lambs is the Horned Dorset, peculiar to the counties of Dorset and Somerset; we, however, sometimes meet with flocks of the same breed without horns, but they are quite an exception, and were originally propagated from the same stock. Early lambs are occasionally obtained from the Southdowns and other breeds, and after many years' futile attempts to obtain the early lambs as a rule from these breeds, it is now considered quite a hopeless case, and the Horned Dorset is the only breed which can be depended upon for that purpose.

In selecting ewes of the horned variety, it is requisite that they should have been put to a Southdown tup, by all means chosen of good quality, being well-made, short-legged, and clothed with fine wool. The Horned Dorset breed of sheep has been greatly improved within the last twenty years; but the number of flocks has been much diminished, having given place to the Southdowns upon the hill farms and in exposed situations. Formerly many flocks of these horned ewes were propagated almost

* "It is worthy of note that sheep have been dipped," says Finlay Dun, "in solutions containing double, treble, and even four times the quantity of arsenic contained in Elliott's mixture, the immersion being purposely continued over three minutes, and yet no harm has ensued. We are inclined to consider the Australian losses, as recorded above, were not the result of absorption, but the consequence of the animals dripping or draining on the food."

entirely with regard to their milking qualities and propensity to produce twin lambs, in doing which the shape of the animal was comparatively disregarded. We still meet with flocks reared in the same manner at the present day; hence the necessity of the before-named selection. During the last twenty-six years I have continued to keep this breed of sheep, and I have found in some seasons, when my ewes have been ill-shaped, that they have yeaned an immense number of lambs, and have proved very milky, and made lambs of the first quality; they would not, however, fatten whilst suckling their lambs. At other times, when I have obtained the choicest description of horned ewes from the best districts of Somersetshire, I have found that they not only brought a large number of lambs, but the ewes and lambs would both become fat and fit for sale at the same time, and in first-rate condition.

These ewes are sometimes sold in the spring of the year; but the usual period at which the breeders of this kind of stock offer them for sale is at Michaelmas, just before the time of lambing. Since, however, the number of flocks have been diminished, they have become comparatively dear, and it is therefore a common practice for some graziers to purchase and keep over for breeding purposes, the second year, those ewes which may, from circumstances, be found poor, or in merely stock condition in the spring of the year. The plan of breeding from the ewes the second season is found to answer a good purpose upon small arable farms having but little pasture land attached—it being the best policy to keep a stock flock in the summer, and a fatting flock in winter; for it must be borne in mind that this kind of sheep does not fatten readily during the summer months in the enclosed districts of the southern counties, because they feel the annoyance of flies more than most other breeds. The custom is to turn the tup with the ewes the first week in May: a short, fine-woolled sheep should be selected, in order that the offspring, more particularly in the case of twin lambs, which generally require to be kept a longer period, may possess a close coat, it being

Wensleydale Longwool Ewes. The Property of Mr. Matthew Wood, Low Ellington, Masham.

well known that loose hollow wool prejudices the sale of lambs in the live market. The tup should also possess good symmetry and plenty of flesh; a well-bred Hampshire Down I have found better than a pure-bred Sussex Down, for the lambs reared from the latter do not possess a due proportion of lean meat, whereas those produced from the former are highly esteemed by the consumers of the present day, affording, as they do, a well-combined proportion of flesh and fat.

The rams should be shorn about a fortnight previous to being turned among the ewes, and kept in an open shed up to that time, in order that they may gradually become accustomed to the loss of their coats; otherwise, in case of their being turned in the open field when recently shorn, they suffer in health and condition during the night frosts, which often happen in the early part of the month of May. When these ewes are kept entirely for the purpose of producing early lambs, they should never be shorn until the rams are taken from them, which should be done about the 20th of June; in that case the portion of the flock of ewes which proved to be pregnant would finish dropping their lambs about the 14th of November. It is not advisable, in a flock of early stock, to have lambs fall after that period; for in case of ewes lambing later, they do not fatten readily with their lambs by their sides.

The manner of keeping the ewes will have its influence in inducing the ewes to offer to the ram at the earliest time. Although the nature of this breed of sheep will go far in this respect, yet circumstances often arise, such as the state of the weather, to delay the breeding season; yet this may in a great measure be prevented by generous keeping, and by choosing a sheltered situation for feeding them.

Let us now, as illustrating the contrast prevalent between the management of the ewe flocks at the two ends of the kingdom, turn to the case of the Lammermoor sheep farm already quoted in these pages. The following are reports from the district during June:—

"*June 1st.*—The lambing season is now over. Though

there was little or no vegetation till the beginning of May, and very little even then, the weather has been otherwise very good, having had scarcely a shower till the middle of this month. On a hill farm we never have lambs for ewes, there being generally from three to five per cent. of barren ewes. This year the deficiency is about six per cent., and on some farms where the ewes were very low in condition, even greater. The lambs are healthy, and only require growing weather to make them strong and good. The Cheviot and crossed lambs (except the youngest) were castrated about the 10th. The wound is slightly anointed with sweet oil and turpentine to prevent inflammation. The ewe lambs get the stock and age-mark at the same time. It is somewhat remarkable that among 600 hogs which have been wintered without turnips, there should have been only one case of sturdy or water on the brain. Is this unusual occurrence (on this farm at least) to be attributed to the jackets (see p. 37)?* It is generally admitted that cold wet weather is the primary cause of this disease, from shedding the wool along the back, and admitting the cold and wet immediately on the spine. This evil is effectually remedied by the jacket, which does not admit one drop of rain. If this is one of the beneficial results of jacketing, the saving of sheep in one year will nearly provide all the cloth required.

"*June 16th.*—The sheep are now beginning to recover the effects of winter, though we fear there will be a considerable loss of wool, and they will not be in a condition to clip for three weeks yet. The black-faced lambs were cut about ten days ago; these are generally the last operated upon; their horns, if cut sooner, being apt to grow in, and injure their eyes.

"*June 30th.*—We are now engaged with the clipping, having made a beginning on the 24th of this month; the sheep were washed a few days before by swimming them three or four times through a deep pool, by which means

* By a reference to Chapter xxi. it will be seen that "Sturdy" is known definitely to be due to parasitism: the larva of the dog tapeworm, carried to the brain, and there assuming the bladder form.

the wool is made very clean. It is the general rule in this district for neighbours to assist each other at sheep-shearing, arranging matters in such a way that each farm may have its own days. On the 24th inst., we had twelve hands at work, who began at seven o'clock in the morning, and finished about 600 Cheviot ewes and hogs by five o'clock in the evening, at which time they were stopped by rain. On the 26th inst., we had nine hands, and finished 400 in the same time. A boy carries away the fleeces, another puts on the stock-mark with pitch, while two women roll up the fleeces. We shall still have two days' clipping, one in the end of this week and another during the following.

"Having, in former years, found quick-lime strewed in the pens an excellent preventive of the foot-rot, we have had this carefully attended to."

THE SHEEPFOLD is carried out industriously all through the month of June on light soils in the southern counties. The flock fed upon the downs may be folded once or twice to a place during night, at first on the fallows intended for turnips, and afterwards on grass and clover, rape, &c., in fields to be broken up in autumn for wheat.

CHAPTER VI.

THE FLOCK IN SUMMER.

The flock in Summer—Weaning of Lambs—Dipping Sheep—Additional remarks on Sheep-shearing, washing, &c., on the Lammermoor farm—The flock in August—Cheviot Lambs weaned—The flock in September—Purchasing Sheep—Foot-rot—Receipes for cure—The hoose—Curative treatment—Purchase of Ewes and Lambs—The Lammermoor farm.

WHERE folding is the system, it should be followed in July with unremitted diligence: the lands usually fixed on for this purpose are the wheat fallows; but the rule is to fold those first which will be first sown. During this month, fold such fields as are destined for August-sown grasses, of whatever sort, or tares: if the manure is left long before sowing, the benefit reaped by the crop will not be nearly so considerable.

Before this month goes out, the lambs of the flock may be weaned. They are put on aftermath of clover, or other good keeping, while the ewes are for a few days put on the barest pastures.

This is the ordinary time for dipping sheep to destroy ticks, and both ewes and lambs are the better for it. The subject was referred to in our last chapter. Ewes should be kept from their lambs, if not already weaned, for a day after being dipped; and their udders should be wiped and cleansed with a cloth.

We add the following to the remarks made on sheep-shearing. The general price of sheep-shearing in Sussex is 10*s.* a hundred and four quarts of beer per man. Half-a-crown a day, with beer, is paid in addition to a man who winds the wool for eight or ten shearers; and there are two men employed to catch and carry the sheep to the shearers, who have 2*s.* a day apiece. Each shearer is supposed to cut forty sheep a day, and for 320 sheep the cost would be—

	£	*s.*	*d.*
10*s.* per 100, with beer . .	1	18	0
Wool winder . . .	0	3	3
Two men to catch . . .	0	5	6
320 sheep, at 2*s.* 11*d.* a score.	£2	6	9

This is the cost in the case of the small Sussex South-down breed. In Gloucestershire the cost of shearing the large Cotswold sheep will be nearer 5*s.* a score than 3*s.*, which is thus the cost in Sussex.

The following is a report, in illustration of the contrast which climate imposes in the management of sheep, from the Lammermoor sheep farm already quoted in our pages:—

"*July* 23.—Until the beginning of this month it can scarcely be said that we have had summer weather. Frosty nights, and cold withering winds during the day, were the marked features in the register for May and June. Under such adverse circumstances vegetation

WELSH MOUNTAIN RAM, "BRHEMIN CYMRU." The Property of MR. J. JONES, Llandudno, Carnarvon. First Prize, R.A.S.E., Warwick, 1892.
(From Professor Wallace's "Farm Live Stock of Great Britain," by kind permission of Messrs. Crosby, Lockwood & Co.)

has made very slow progress, and there has been a greater want of grass on our hills during the months of May and June than has been the case for some years past. The effect of this on sheep stock is very apparent, more especially on lambs, which have neither the condition nor the blooming skin they had last year. For three weeks, however, we have had genuine summer weather, which has had the effect of magic on the pastures and everything else. We finished the shearing of sheep on the 13th, being fully a week later than usual. This arises partly from the cold weather, and partly from the want of condition having considerably retarded the growth of the wool. When washed, the sheep are driven three or four times through a deep pool, about eighteen yards broad, and clipped three days afterwards. Whenever a lot are clipped, they are driven once through the washing pool, to wash away any dirt that may have adhered to them during the operation. This practice is not a common one, but has long been followed on this farm, and was first adopted on account of the frequent deaths that occurred immediately after clipping. Two years ago it was discontinued for one season, and the consequence was that a considerable number died. Since then it has been regularly done, and there has been no loss. Now that the grass is growing and the ewes want their fleeces, we expect the lambs to improve rapidly. The principal business of the shepherds at this season is to keep them clear of maggots, which in warm, moist weather, are very troublesome. The lambs will be weaned about the middle of August.

THE FLOCK IN AUGUST.—The earlier lamb fairs of the season take place, and the surplus stock of lambs, where breeding flocks are kept, is parted with. Attention is still necessary to guard against the attacks of the fly; and this is the usual season to dip the lambs as a preparation against tick, lice, and scab.

In the south of England the ram is turned with the ewes in August for the main crop of lambs. The horned ewes are now forward in lamb, and commence lambing, in fact, as early as October.

Among the Lammermoors, on the other hand, the following is the report for August:—

"The Cheviot lambs are all weaned on the 10th, and after having been sorted, a sufficient number of ewe lambs are retained to keep up the stock, while the smaller lambs are sold for the purpose of breeding "half-breeds," or crosses between the Cheviot ewe and the Leicester tup. The top wedder lambs are driven to market. The ewes are milked once or twice, to prevent the milk injuring their udders. Of the black-faced lambs, such as are intended for keeping are removed from the ewes about the middle of August, whilst those which are destined for the butcher must remain a little longer, to suit purchasers and the different markets to which they may be sent. Towards the latter end of September the drafting and sale of the oldest ewes will take place. The Cheviots are disposed of at four, and the black-faces at four and five years old."

September.—The ram is now put with the ewes—one to sixty or seventy—even in the later districts of England, and both are put in good pastures, and kept in condition by ample feeding.

This is the season when it is proper for those who do not breed their own stock to be purchasing for winter feeding. Early turnips may be commenced with sheep. Such a stock should be purchased as shall consume all the white turnips by the middle of November, the yellow sorts by the end of the year, the Swedes by the end of March, and the mangold wurzel by May. And to assist calculation, it may be assumed that good crops of the above roots on ordinary soils will yield per acre 24, 20, 16, and 26 or 30 tons of clean roots respectively; and that a fatting animal, either sheep or ox, supposing it full-grown, will in general, when out of doors, eat a weight of these roots equal to about a quarter of the carcase weight it is expected to attain when ready for the butcher. If fed under shelter, three-fourths of this quantity may be assumed as the basis of calculation. If this be true, and from a somewhat extended experience on this subject we believe it to be so, what a saving does this shed-feeding hold out to the winter grazier! It is objected to the system that it entails the heavy expense of carting the crop to the buildings, and the dung back again; and, in the case of sheep, that it renders them especially liable to the foot-rot; but the first item may be greatly diminished by erecting a temporary shed on the neighbouring stubble-field, which will require all the manure made there for the ensuing green crop, and thus the second item will disappear. Such a shed may be easily erected with poles and hurdles. A few larch poles, nails, hurdles, and bundles of straw are the materials required; they are not costly, and certainly they will not exceed in expense 6*d.* for each of the sheep they shelter. And as regards the foot-rot,* it need not be troublesome if you pare the overlapping horn from the hoof once at least every month, and apply a light caustic ointment on the first symptoms of lameness.

The following is a recipe by Mr. W. C. Spooner, V.S., of Southampton. The cure consists in removing sufficient horn to allow any confined matter to escape, and to apply a styptic or caustic to the part, so as to prevent proud flesh growing, and to stimulate the vascular parts to secrete healthy horn. A number of medicines have been recommended, and with success: muriate of antimony applied with a feather is a very convenient caustic, and so are likewise equal parts of hydrochloric acid and tincture of myrrh; a strong solution of sulphate of copper is a milder dressing, and tincture of aloes or friar's balsam is milder still.

All these are useful if applied with discretion, according to the severity of the case. Another useful application, as well as a preventive, is coal tar, particularly if a little creosote is left in or added to it; and some powdered plaster of Paris over the tar will assist its drying effect. A change to a dry pasture or soil is necessary. Sulphate of lime, powdered, 1 oz.; sulphate of zinc, powdered, 1 oz.; creosote, 1 scruple; Stockholm tar, 4 oz.; lard, 4 oz.; to be made into an ointment, and applied occasionally to the feet of the sheep and between the claws,

* See Chapter xxiv.

first paring the overlapping horn away, and clearing everything away without drawing blood, is a good recipe.

Lambs are at this season sometimes troubled, as calves are, with the hoose,* a disease whose character is sufficiently attested by the *post-mortem* examination of its victims revealing the presence of an immense number of small worms in the windpipe and bronchial tubes of the lungs, so that we cannot wonder at the gradual pining away which denotes, or the fatal result which so often follows, this affection, if allowed to go on unchecked. Not only is the body deprived of its nourishment by these innumerable parasites, but the proper changes of the blood are retarded or prevented from being accomplished in the lungs.

The following are remarks by Mr. Spooner on this disease:—"Any medicine given to destroy worms in the air-passages acts by being absorbed into the system. In the case of worms in the windpipe, two objects should be sought for—one to destroy the enemy, and the other to strengthen and support the system which is being subjected to such debilitating and exhausting influences. In the case of the 'gapes' in chickens, a similar affection, some speak of the good effects of tobacco-smoke introduced so as to almost produce suffocation, albeit the creatures sometimes die of the remedy. In calves a cure has been effected by administering lime-water, and probably it would be equally effectual for lambs. The dose for a lamb would be about two ounces daily, and about two drachms of salt should be given at another portion of the day. This treatment should be followed for some days. Better still is the plan of administering oil of turpentine, which being taken into the stomach, is soon absorbed throughout the system. The dose for a lamb is two drachms, which should be given with an ounce of linseed oil, a scruple of ginger, and five drops of oil of carraways, mixed up with two or three table-spoonfuls of linseed gruel. This dose may be repeated if required several times, with intervals of some days. The lambs should be allowed half a pound of linseed cake per diem,

* See Chapter xxi.

Blackfaced (Scotch) Mountain Ram, "Lochiel." The Property of Mr. C. Howatson, Glenbuck. First Prize at H. & A. S., Inverness, as a Shearling, 1892; First at H & A. S., Edinburgh, as Wool Sheep, 1893, and Second as a Two Shear.

and should be otherwise carefully tended and liberally fed. By such a course of treatment many valuable animals may be saved."

There are, perhaps, few diseases which have so well resisted the action of various remedies, as the one under consideration. In their selection too much attention has been concentrated upon the likelihood to reach and destroy the worms, their effect upon the constitution being altogether overlooked. It is not, therefore, surprising that thousands of lambs succumbed to the remedies rather than the disease, the mortality being attributed to the wrong cause. Nevertheless the losses were otherwise serious, and continue thus at the present day, for the obvious reason that the vitality of the parasite is simply interminable under certain conditions. The preferable mode of treatment consists of the injection of remedies within the windpipe, by piercing the spaces between the rings with a suitable syringe armed with a hollow needle. A mixture of "Sanitas" oil, chloroform, salicylic acid, and glycerine often proves satisfactory.

Tegs put to fatten in the fold on turnips or rape should have clover hay given in racks along with the usual fold of green food. And those which are forward in their fatting, and are to be sold at Christmas-time, may receive the best food which it is the practice to give—as 1 lb. of oil-cake or 1 pint of peas daily apiece.

When it is the custom to buy ewes for the sake of a crop of lambs, they should be purchased early in September. Young ewes are to be preferred, and it is well to choose a ram of a larger breed than themselves, as a Cotswold ram to a lot of Down ewes. This is in the case, now not uncommonly adopted, of lambing down a lot of ewes, and fatting both lambs and ewes off as rapidly as possible in the following summer.

We add here the monthly notice of proceedings on the Lammermoor sheep farm :—

"We commenced cutting corn on the 30th August, and finished on the 11th September, and expect, if the weather continues favourable, to carry all this week. This is a very early harvest. We have still some hay to stack, the weather having been so windy that it has frequently been impossible to meddle with it. It is all, however, in large ricks in excellent condition. The carting of the hay is a tedious business, most of it having to be taken several miles away over very bad roads. It is put up in stacks of 300 or 400 stones each, beside the best sheltered stalls; this quantity ought to keep 200 sheep for a month, provided they are able to scrape something from under the snow. The lambs are now all away from the ewes, and nearly all sold, except the ewe lambs required to keep up the stock. During the past summer we have been very careful in having the sheep-pens well strewed with quick-lime, whenever we had occasion to have sheep in them; the consequence is, that we have scarcely any foot-rot. The draft ewes will be marketed forthwith."

CHAPTER VII.

THE SHEEPFOLD IN AUTUMN.

The sheepfold in October—Treatment of Tegs—Shearlings fattened for Christmas—The Lammermoor farm—Bathing Sheep—The sheepfold in November—Lammermoor farm—The flock in December—Value of white carrot—Lammermoor memoranda—How to estimate the weight of Sheep—Food and increase.

THE earlier horned ewes in the southern counties, where house lamb is fed for Christmas, are now dropping their lambs. Reference has already been made to this exceptional practice.

The rams are not yet taken from among the ewes in ordinary farm practice; and on the moorland farms of Scotland they are not generally admitted to the ewes until November.

Tegs (lambs) which have been on clovers, &c., during the summer, are put on the earlier turnips and rape this month, and folded there, receiving in addition to their daily food some hay chaff; and when the object is to bring them out ready for the butcher immediately after taking their coats in spring, they should be gradually brought to some richer feeding by the addition of peas, or barley, or oil-cake, whichever may be the cheapest—beginning with less than ½ lb. a day apiece. Shearlings which may be fattening for Christmas-time, or thereabouts, should receive either in the turnip fold or in the yard, as already directed, a larger allowance of corn or cake with their green food.

All this is the practice of the lowland districts. Among the moorlands where sheep husbandry is confined to the breeding of the black-faced or Cheviot sheep, a very different practice prevails, of which the following report from the Lammermoor farm is a sufficient illustration:—

"*October 10th.*—We are now in the midst of harvest, and rain and snow have this day been falling. Cutting is not yet finished in the district, and much grain was shaken by the wind of last Wednesday. On the earlier farms a good deal of corn has been stacked, but on the higher lying and later ones little more than a beginning has been made. Turnips look well, and every endeavour will be made to secure them against frost, either by storing or earthing up in the drill.

"Our hill flocks have now again in a great measure regained their condition, as the summer has been a good one for hill pastures, and keep abundant. But we question very much how they would stand a second bad winter. Our stock of rams is provided, and we have taken turnips for them elsewhere, that they may be out of the way of mischief, and be in good condition when they are required, about Martinmas (Nov. 11). The bathing of the sheep will commence in about ten days, and will occupy our shepherds for a fortnight. This year we intend using the turpentine bath, a preparation which both improves the quality of the wool and kills all vermin. The cost is about 2½*d.* per sheep."

Bathing the sheep is an operation for the purpose of killing the sheep ticks and of meeting any tendency to cutaneous eruption. The bath is generally a decoction of tobacco mixed with spirit of tar and soap, and it is poured in upon the skin of the sheep along seams of parted wool, two or three inches apart, so as to thoroughly wet the whole surface of the skin.

As a remedy for the cure of scab, destroying lice, ticks, and maggots, the preparation known as "Sanitas" deserves an extended trial. It has the valuable property of being non-poisonous to the sheep and higher animals, even when applied persistently and in any quantity, but is eminently fatal to the lower organisations of parasitic life. It leaves the wool soft, perfectly clean, and white, and the fly will not strike as long as the remedy remains in the fibre of the wool.

The preparations for dipping sheep have been numerous and in most cases the remedy has proved worse than the disease, creating wholesale destruction among the flocks by absorption of the poison through the skin. In this way many preparations from coal-tar, containing carbolic acid, were almost as fatal as the arsenical solutions, though the latter added much to the injury by the poisoning of the herbage on which the animals were allowed to drip. "Sanitas" is free from all these objections.

THE SHEEPFOLD IN NOVEMBER.—The ram is generally removed from among the ewes in lowland districts towards the end of this month.

The practice of yard feeding of sheep has already been described. It may in general be profitably carried out.

Fatting sheep will receive either a fresh fold in the turnip-field, along with hay chaff, and a pound of cake, or peas, or other corn; or cut turnips, in troughs in the field or yard, along with corn, or pulped roots and chaff and meal mixed.

The following is the report of the present date from the Lammermoor farm:—

"*Nov. 24th.*—In high districts such as this, grass grows little or none at this season of the year, still the pastures are tolerably fresh, and sheep continue to do well. The heather is now in perfection, and forms more than half of their food.

The bathing of the sheep is now completed, having occupied us about fourteen days. Five shepherds and

BLACKFACED (SCOTCH) MOUNTAIN RAM, "STIRLING." The Property of MR. C. HOWATSON, Glenbuck. Second Prize at H. & A. S., Inverness, as a Shearling, 1892; First at H. & A. S., Edinburgh, as a Two Shear, 1893, and First as Best Sheep of the Breed.

five boys or girls employed; the shepherds shedding the wool, and the others pouring on the bath. Each man does about sixty per day. The tups were put to the ewes on the 23rd. We allow fifty ewes to one tup. This is a small number, but from the extent of ground they range over it is unsafe to give more. The usual period of gestation is about twenty-one or twenty-two weeks, and the tupping season does not commence in the Highland moors till now, lest the mouths should arrive in spring before the food for them."

THE FLOCK IN DECEMBER.—The fattening sheep continue, whether in the fold or in the yard, to receive the same good food and treatment already described. Those in the turnip fold receive a daily allowance of 1 lb. of corn, with as many turnips and as much chaff as they can consume.

The earlier breeding flocks begin to drop their lambs towards the end of the year; and as already said of the Hampshire farm, the horned ewes have lambs now many weeks old.

Mr. Blundell, of Southampton, speaks very highly of the Belgian carrot as food for them. He says:—

"Having noticed the great value of the white carrot for the feeding of horses and pigs, I was induced to commence an experiment for the purpose of proving the value of this root for the feeding of early lambs, as compared with that of the Swedish turnip. I began with giving the carrots, cut with Gardner's machine, in troughs placed side by side with others containing Swedish turnips, cut with the same machine, in advance of the hurdles, in the usual manner; and I found, in the course of a day or two, that the lambs had so great a preference for the carrots that they ate no Swedes so long as any carrots remained in the troughs; in consequence, I discontinued giving any cut Swedes whilst I had any carrots for use, they having continued in good condition until the 1st of April.

"Finding my early Somerset lambs doing so well upon carrots, I decided upon feeding the Southdown lambs in the same manner, and the result was precisely the same, they having refused to eat any cut Swedes when they could get carrots. I had no means of proving with accuracy the difference in the quantity of carrots and Swedes consumed by a given number of lambs, neither did I attempt it, considering their great liking for the carrot quite conclusive; although I believe a lot of lambs would consume a greater weight of Swedes than of carrots. The quantity of hay consumed was about the same as in former seasons, when feeding with Swedish turnips. But I was struck with the apparently diminished quantity of corn and cake consumed by the lambs whilst feeding on carrots; I therefore determined upon ascertaining the actual quantity of oil-cake and peas consumed by a lot of 100 lambs whilst feeding on carrots, as compared with former seasons when feeding on Swedes; and having in previous years found that 100 lambs, being allowed as much as they could eat, consumed on the average four gallons of peas and 26 lbs. of oil-cake per day, for a period of nine weeks, commencing at five weeks old, until they were fourteen weeks old, being at that age fit for sale; I then found, to my surprise and satisfaction, that 100 lambs, being fed *ad libitum*, only consumed two gallons of peas and 14 lbs. of oil-cake per day (having at the same time the usual allowance of hay and as many carrots as they would eat), or, in fact, about one-half the quantity usually consumed before commencing the experiment of carrot feeding. The advantages to be derived from feeding lambs with carrots, I find, consist in the saving of one-half the cost of oil-cake and corn, and in the lambs being fit for the butcher earlier, and attaining a greater weight and better quality at a given age, than when fed on any other root. In proof of which, I have never, during a period of more than twenty years, sold lambs so fat and heavy at the age as those which I have fed during the progress of this experiment, and in the manner here described."

The treatment of fatting early lambs has been already described.

In the moorlands of the north the ewe flock are treated

very differently. The following is the report from the Lammermoor farm:—

"The bathing of the sheep was completed on the 17th November, and the jackets were all got on the hogs by the 22nd, on which day the rams were turned out to the ewes. As we do not hirsel (arrange in separate flocks) our sheep, but allow them to distribute themselves equally over the ground, we only allow forty-five or fifty ewes to each ram. With more than this number, barren ewes are apt to be too numerous. We think it better not to put a ram to the same ewes more than one season, experience showing that the first cross takes most to the sire, and the second to the dam; and as the former is supposed to have been well selected, and to possess at least some properties superior to the ewes, it is well to avoid whatever may retard the general improvement of the stock; especially when it can be so easily avoided, even on our open ground, by the shepherd putting a sheep to those ewes which graze the opposite side from where he was the previous season. As the progeny of a shearling ram are considerably more vigorous than those of a three or four-shear, we endeavour to have the majority of our number of the former age, though a really superior sheep will never be laid aside until he is inefficient from age. Such of the cast ewes as were unfit to be sold for breeding are getting turnips, and about half a pound of oats daily, with the view of fattening them, and getting them off as quickly as possible. During severe weather their allowance of oats will be increased."

The weight of the carcase is to the live weight of an animal as 8 to 14 or thereabouts in fat sheep. But these proportions vary according to the condition and breed of the animal.

Food and Increase.—Sheep fattening on good food composed of a moderate proportion of cake or corn, hay or straw chaff, with roots and other succulent food, will consume about 15 lbs. of the dry substance of the mixed foods per 100 lbs. live-weight per week; and should yield, over a considerable period of time, one part of increase in live-weight for about nine parts of the dry substance of their food. If the food be of good quality, sheep may give a maximum increase even provided the food contain as much as five parts of total non-nitrogenous* to one of nitrogenous compounds.

Moderately fattened sheep (shorn) should yield about 58 per cent. carcase in fasted live-weight; excessively fat sheep may yield 64 per cent., or more. The proportions will, however, vary much, according to breed, age, and condition.

Of the *increase* over the final six months of liberal feeding, of moderately fat (one and a quarter to one and a half year old) sheep, 65 to 70 per cent. may be reckoned as saleable carcase. Of the *increase* over the final six months of liberal feeding, of very fat (one and three-quarters to two years old) sheep, 75 to 80 per cent. may be reckoned as saleable carcase. Sheep, fattening for the butcher on a good mixed diet, should give about nine parts dry increase—consisting of about eight parts fat, 0·8 to 0·9 part nitrogenous substance, and about 0·2 part mineral matter—for 100 parts total dry substance consumed. More than ninety parts of the consumed dry substance are, therefore, expired, perspired, or voided.

* The cereal grains contain on the average rather more than six parts of total non-nitrogenous to one of nitrogenous compounds; and the leguminous seeds often not much more than two parts to one. Oil-cakes and foreign corn contain rather more than six-sevenths, and homegrown corn, hay, &c., rather less than six-sevenths, of their weight of "dry substance." Common turnips generally contain about one-twelfth; Swedes about one-ninth; mangolds about one-eighth; and potatoes about one-fourth of their weight of "dry substance."

CHAPTER VIII.

DISEASES OF SHEEP.

General observations—Value of our domestic animals—Public health dependent on the health of stock—Study of Sheep diseases frustrated—Opportunities offered—Appeal—Losses, and how they may be averted—Pathology—Symptoms—Morbid pathology—Veterinary medicine—Veterinary surgery—Materia Medica—Fever—Inflammation—Abscess—Serous cyst.

THERE are, perhaps, few departments in our social economy which have exhibited a greater display of patient perseverance and high intelligence than that engaged in the improvement of our domestic animals. The British stock-owner claims just pride in his noble and majestic horses, the purity of his shorthorns, and the renown of his improved flocks. Their almost absolute perfection is the reward for sagacity and profound veneration for the pursuit he has selected; and by his devotion to it, combined with a prescience descending to him as a national heritage, the state of English agriculture is such as to influence both present and future prospects of our commerce and industries generally.

We cannot, however, record the same progress or elaboration with respect to crops and land. By reason of the peculiar conditions with which they are associated, and the laws which govern and control them, an acquaintance with the most intricate sciences is necessary towards that end; but this, hitherto, has not been comprised in the education of the agriculturist. Mere observation and acute natural intelligence, which have done so much for the improvement of stock, fail to do the same for the land and its crops; yet he is little behind his compeers in other and more highly favoured countries in this respect. Soil, under the varied influences of climate, locality, drainage, and moisture, as well as constitution, is capable of generating in the crops a nature which may not only render them valueless, but highly injurious as food. By the inelastic spirit of the covenants with which the farmer is bound, he dare not exercise his intelligence with the view of improvement. This and a host of other causes, as excessive rents, the burden of taxes, state and rate of foreign importations, &c., &c., have a material effect in crushing his spirit; and these, as comparatively modern evils, have stepped in to connect the past age with the present.

In the care and preservation of that unequalled stock, the agriculturist has received but little from without. That innate love of perseverance which has always characterized him has caused him to plod along contentedly, though alone. The want of a higher, if not a scientific, education is not felt where observation alone will serve the purpose for the time being; he, therefore, failed to see how much the cultivation of sound knowledge, in reference to the diseases of animals, could aid him either in a simple ailment or in the calamity of an epizoötic. As far as they were capable, his observation and intelligence served him in this particular, and when they failed the menders of boots and welders of iron were appealed to, as having at least as much information on these matters as himself. As the barber once ministered to the ailments of mankind with as much skill as he cut the hair or shaved the chin, so the cobbler and smith would be sure to possess a fund of information on the diseases of stock, for no other reason than that the former shod the master and the latter shod the master's horse.

Beyond the bare fact that cattle and sheep are grown for sale, the farmer's coffers being enriched thereby, we must take into account the additional circumstance that they are grown too for human food, and the coffers of the farmer can only be permanently lined by the quality of the production. In other words, the health of the stock is of paramount importance. If it is not good, the quality of the production is inferior; the public health suffers, the commodity is shunned for a season, and the credit of the producer is at stake. It may be thought, and indeed it generally is, that the killing of animals in the throes of death is the most orthodox and humane method, and the subsequent sending of the flesh to the nearest town for disposal equally justifiable. But the public are now beginning to judge the proceeding as not only dishonest, but criminal—a most unjustifiable attempt to injure the public health.

Hitherto the relationship existing between the health of stock and that of the human population has not been recognised: how much less understood. The dependence of animals upon each other for their share of freedom from disease is also a theme which must claim greater respect and attention than it has yet done. In both these departments there is a call for the exercise of skill and sound judgment. There is need for the adoption of principles which shall specially apply to the preservation of the health of our stock, and thus secure sound flesh as human food; and the necessity for these once rightly understood by the farmer, he will find his profits much larger and more permanent than by attempting to make marketable the disreputable carcases which often find their way to the tables of the poor. It is a libel upon the man of intelligence and presumed honesty to say such proceedings are worth his attention. Let him study the position which he occupies in relation to the public good, and the proud spirit of his forefathers will kindle a revolt in his nature. Like his corn and other commodities, his cattle should be grown and sold in good condition. There is no reason why the losses on a farm should not be reduced to the smallest minimum; and this may be done by taking advantage of the assistance which veterinary science can offer.

There are, perhaps, no other domestic animals which have received less attention at the hands of the veterinary practitioner than our immense flocks of sheep. It has been said, and truly, that veterinary surgeons have not cared to undertake the treatment of the ailments of these animals. This is true in a number of instances; some, indeed, have not cared even to study them, knowing full well that their services as country practitioners would probably never be required, or at least seldom, and only as subordinate to the opinion of a shepherd. Under these circumstances so little common sense, and less scientific skill, has been brought to bear upon this important subject, that our flocks are decimated year by year by maladies both curable and preventible. The sheep of our hills are capable of being produced to an enormous extent, and to afford a corresponding yield of profit; their value indicates the necessity for closer supervision, in order that it may not be jeopardised; and, as their carcases form such a large proportion of human food, we have an additional reason that care should be exercised to insure the flesh being all that is desired in point of purity and excellence. We urge, therefore, a closer union between agriculture and veterinary science. The haunts of the sheep on the hills are capable of affording immense scope for observation and the accumulation of facts, which a trained mind will know how to utilise for the good of the flock and profit of the owner; but as long as the distance is maintained, so long must the value of the untried services be misunderstood. The wealth of the nation largely depends upon the health of our cattle and sheep especially. There are numerous diseases capable of being transmitted to mankind from both. Such diseases cannot be understood in their nature, origin, cause, cure, or prevention by any but those who have had a special training in medical science; therefore, as long as the health of the flock is in the keeping of those unqualified, so long must the public health suffer, and by devastating contagious and other preventible maladies the breeders' profits are sapped. Many thousands of pounds are annually squandered in worthless drugs—worthless as remedies in numberless instances because they are wrongly applied, but they fail not to exercise an effect the reverse of good. All this might be readily saved under proper medical supervision, besides the lives so frequently sacrificed in addition. The health of our flocks and herds means so much of health in the human population. Let the farmer grow rich by his success in breeding and rearing them, for his riches bear a corresponding ratio to the comfort and satisfaction of the people who derive a wholesome food at a fair price. Besides this, the prosperity of the farmer is the country's good. Agriculture exerts an important influence on our commerce and home industries, and no country has so bitterly acquired this experience as the people of Great Britain during the past forty years.

The mission of veterinary science, as usually understood, is entirely that of curing the maladies common to the lower animals. The limit thus placed upon it has been fatal to the security of the farmer, and also to the progress of the science. While we admit that curing disease forms a part of the duties of the veterinarian, it is obvious that he can only do so after fully understanding the causes, nature, symptoms, pathology, and morbid signs of each as it arises; but the aim and scope of the science is by the same means to point out the principles of *prevention* as well, and this should be paramount to every other. Many diseases are yet not perfectly understood. The apathy and neglect prevailing in many places stand in the way of investigation, and obscure causes and signs are not fully made out. They are altogether unrecognised in the fold and on the hills. Thus many affections are unnoticed, especially those of a contagious or epizoötic nature, until scores of lives are sacrificed, and possibly contagion is carried to the flocks of intelligent owners who take every ordinary precaution, but under these circumstances are caused to suffer in common with those who, caring little for their own, perhaps value likewise the property of others.

That department of veterinary science which takes cognisance of the peculiarities of disease—its special characteristics, which point out its locality, nature, &c., we term *Pathology*. Then, as to various diseases, we recognise certain indications, such as increased, decreased, or arrested action of various organs, as a result of morbid action. The physiology of health is modified, and the manifestations it gives rise to are called signs, or *symptoms*, and these are of such a nature as to be said to belong to one particular disease or a class of diseases. Symptoms are faithful monitors, which direct our attention to particular points during life, and by which we judge of the existence or non-existence of morbid action. In addition to symptoms, the practitioner calls to his aid after death morbid signs, or the state to which the tissues have been brought by the action of disease. The standard by which he judges is the healthy condition of the organs, furnished in a previous course of anatomical study, and the evidences obtained are those which confirm the language of the signs observed during life. This is known as *Morbid Pathology*, and is described in veterinary works under *post-mortem appearances*.

By the term *Veterinary Medicine*, we comprehend the enumeration and consideration in detail of the nature, causes, symptoms, and morbid appearances of all internal diseases. *Veterinary Surgery*, on the other hand, implies a similar acquaintance with external affections, as well as those resulting from accidents, operations, &c. There is, however, no sharp line between them. *Materia Medica* is that department which describes the physical character, nature, action, doses, forms of combination, prescription, and administration, &c., of the various remedies. The list is not only great, but many are of special character, and suitable only to particular animals and certain diseases. From this circumstance, to say nothing of others, the practice of veterinary medicine calls for the closest observation, exercise of skill, and the nicest discrimination. It is not, therefore, an occupation which any or all may take up and utilise when convenient, as one may take up a fork or a broom. The supposed simplicity of dispensing and administering drugs has led to a wrong estimate of the nature of disease generally ; it is believed to be some material obstruction blocking up a simple tube—the system—and all that is required is a violent remedy to clear it out. It is not, therefore, surprising when, after such generally wholesale measures, the affected animal is also cleared out. With so much ignorance among those who attend upon animals, accidents are not to be regarded as wonderful; we, on the contrary, are rather surprised they are not more numerous.

For a more detailed account of the diseases of sheep,

the reader is referred to the larger work, "The Cattle Doctor," by the author of this manual, published by F. Warne and Co., 15, Bedford Street, Strand, London, W.C., and may be obtained through any bookseller in the United Kingdom or British Colonies.

Pathology, or the Conditions of Disease.

Throughout subsequent chapters will be found an enumeration and brief consideration of the various conditions which constitute the acknowledged diseases of sheep. Notwithstanding the many difficulties standing in the way, the pathology of ovine maladies are now much better understood than formerly. Greater facilities have been offered, and the list once made up of the most outlandish affections, having cabalistic terms which connect their history and origin with the witches, dwarfs, and fairies of past centuries, has been revised. It has been pruned of its mystical characters, and considerably augmented by the addition of ascertained facts as to the existence of distinct and special forms of disease.

We do not feel called upon to render a perfect classification or grouping of diseases. This is not of great moment to the general reader. We shall, therefore, in preference, observe a categorical order, and thus endeavour to afford facility of reference to those who run as they read. As a preliminary to the context on general diseases, we will conclude this chapter with brief remarks on *Fever, Inflammation, &c.*

Fever.

By the term fever we understand a disturbed condition of the system, in which the major part, if not all, the secretions are deranged or altogether withheld. The nervous and circulatory systems, together with the organs of respiration, reproduction, and notably those engaged in nutrition and depuration, are more or less involved, or their functions suspended. Such being the state of affairs, the system is bordering upon the development of some form of more serious derangement, which, if permitted to proceed, becomes a confirmed condition of disease.

Fever is understood to be of two kinds, using very general terms, viz., *Simple*, or *Ephemeral; Sympathetic*, or *Symptomatic;* and *Specific*.

Simple Fever.

Is defined to be that shortlived or ephemeral kind of disturbance which may be noticed in highly fed animals, occupying close folds or buildings, and breathing a warm, but not necessarily a tainted, atmosphere. A slight increase of circulation, respiration, and animal temperature will be observed, together with a dry muzzle, diminished secretions, and slight constipation. In a few hours, perhaps not more than one or two, especially after a draught of water, or moderate meal of roots, green food, &c., all have disappeared, and the animal seems none the worse. We cannot, however, close the understanding to the belief that this state of so-called ephemeral or simple fever is not without significance; and, among all animals receiving a highly nutritious food, perhaps confined indoors, and taking little or no exercise, the origin should be traced, with the view of prevention as well as dissipating it, by means of suitable food of a cooling and laxative nature, and a necessary amount of exercise.

Sympathetic or Symptomatic Fever.

May be regarded as a violent aggravation of the simple or ephemeral kind. A few illustrations will serve to make our acquaintance with it quite clear. All minor cases of irritation, as simple scratches, or friction of the skin, a slight chill, &c., may give rise to nothing more than simple fever; but the plunging of a nail into the foot or a muscular part, the fracture of a leg, laceration of the skin and muscles, acute inflammation of some important organ, &c., &c., will occasion such a vast disturbance that the life of the animal may be sacrificed. The immediate cause of death may not be the actual disease, but sympathetic fever, a state which we define and speak of as *constitutional disturbance*, due to sympathy with diseased or injured parts; and such signs being common or pretty general, excepting, of course, the specific manifestations of injury, it is said to be symptomatic of the original disease.

Symptomatic fever doubtless arises by means of the nervous system, which brings all parts into union and sympathy. It is Nature's telegraphy, by which the disease of outlying members is made known in the central system, and the signs are indicative of disturbance in that region of an undoubted and urgent kind. The respiration is accelerated, and it may be *blowing;* the pulse betokens a surprising activity, and is said to be *full and bounding;* arrest of secretion from disturbed function has diminished secretion of urine, which possesses a very *strong odour* and a *deep colour;* the bowels are restrained by the same cause, and the stools are *hard, dry*, and *glazed* with *mucus*. The muzzle is dry and hot; the horns, ears, skin, legs, and internal part of the rectum, or gut, are likewise hot; the wool becomes harsh and clapped; mucous membranes are reddened, and the countenance assumes an expression of anxiety as pain increases; external parts afterwards assume a variable temperature, and at length are cold. The disturbance has been chiefly centered in the lungs, which have become congested; they cease to fulfil their functions; blood is no longer purified and fails to afford nutrition; the nervous system can bear up no longer; and, these states becoming aggravated, the creature dies—not as a result of the original injury or disease alone, but from symptomatic disturbance in other organs, mainly indicated by congestion or stoppage of blood in the lungs. The veterinary practitioner, recognising the train of symptoms developed under these circumstances, views them as a whole, and, in alluding to them either orally, in writing, or even mentally, does so in general terms. They are so well known as collective signs of some serious condition that he ceases to separate them, and at once speaks, writes, and thinks of them as *sympathetic* or *symptomatic fever* and *constitutional disturbance.*

Specific Fever

Is another general term for peculiar and important states. It need not occupy our attention at the present further than to point out that it is applied to the many kinds of disease which form serious foreign contagious as well as the indigenous enzoötic and putrid diseases. Sometimes it is typhoid or typhus in character; at others, benignant, yet prevails as a tardy enzoötic; and occasionally it may be rheumatic. We shall refer to it more in detail as the signs of these diseases are enumerated.

Treatment of Fever.—As fever arises from some already existing source of irritation, it is obvious that the first step towards a cure consists in removing that cause, when Nature re-asserts her sway. The measures adopted entirely depend upon the situation and peculiarities of the parts affected; therefore, when the original disease is controlled, we reduce also the fever which that disease gave rise to.

Inflammation.

The infallible signs of inflammation are *heat*, *pain*, *redness*, and *swelling*. The first and second are usually evident; but, owing to the existence of hair and other coverings common to the animal body, redness and swelling are not so readily observed. Inflammation is the result of violent causes, and consists of an increase of blood in the part affected, with more or less suspended function of the blood vessels, as well as the integral parts of the blood itself. Inflammation is said to be *acute* when the process is characterized by great activity; it is *atonic*, or *subacute*, when, by reason of low vital force, it proves slow or tardy. Closely allied to this state is a peculiarity of the circulation, mostly common to large organs with abundant vessels and elastic tissue composing their substance, known as *congestion*. It is sudden in its origin and departure, and, probably, is confined to the venous system.

Inflammation terminates in various ways: by *resolution*, or gradual decline, the parts eventually regaining their original state and appearance; in *suppuration*, or the formation of an *abscess*, or sac containing *pus*; in *effusion* from the surface of membranes, as water (serum) or mucus; or by lymph within the inflamed structures, by which permanent enlargement or thickening may be the result.

Inflammation is further distinguished by the structures it attacks. Thus we have *serous* and *mucous* inflammation, the serous and mucous membranes being affected. When located in the substance of organs, it is called *parenchymatous*; if it seizes fibrous structure, as the coverings of joints, ligaments, tendon, &c., it is *rheumatic*; and inflammation of the skin and deeper-seated tissues is termed *erysipelatous*. Beyond these terms we need not pursue the description.

Treatment of Inflammation will be noticed under the several diseases in which it forms a special feature.

Abscess.

The formation of pus, more commonly known as matter, among the soft parts of the body, is known by the term abscess. The signs are swelling and heat, with unusual tenderness; and, as the abscess becomes complete, the hair is removed from the central or highest portion, which afterwards is moist. This part is also acutely inflamed and remarkably sensitive, and by degrees becomes thin; movement of the fluid within is readily perceived by pressure from the finger, and the central portion, after bulging outwards for some time, finally gives way under ulceration, and pus escapes naturally. In order to avoid this delay, and preserve the animal from suffering, as well as to facilitate the subsequent recovery, the surgeon opens the abscess by means of a bistoury, or, more frequently, a lancet.

Treatment.—Favourable progress of an abscess is betokened by the signs we have already given, and is treated by the surgical operation as soon as pus is fully formed. Cleanliness, and occasionally poultices and fomentations, are needed; but, as a rule, the first is sufficient. Tardy or slow abscesses may need constant fomentations, poultices, or even blisters. In these states it is sometimes difficult to cause the process of pus formation to go on properly. Great care is then needed, as diffused abscess or a suppurative state may follow, in which abscesses form in various parts of the body, and the animal, if he lives, proves worthless. Good food, healthy habitations, and tonic medicines are particularly called for.

Occasionally the abscess will be found a great depth below the surface. In such cases considerable skill is required in order to reach and liberate the pus without destroying surrounding parts, or causing the death of the animal. The discovery of a deep-seated abscess is often a difficult affair, its presence seldom or never being apparent to the non-professional observer.

Serous Cyst,

Sometimes called the *serous abscess*, is occasionally seen among sheep, particularly about the knees. It consists of a soft, fluctuating tumour, without evidences of heat or tenderness, as a rule. The causes are blows or falls, when the injury is confined chiefly to the skin. The cavity contains a thin fluid, coloured by an admixture of blood, surrounded by an expansion of condensed cellular tissue, covered outwardly by the skin.

Treatment consists in liberating the contents by means of a suitable instrument, and applying pressure, healing fluid, &c. Care is required to distinguish between the serous cyst and bursal enlargements. If the latter is opened in mistake for the serous cyst, the results may be lamentable—violent inflammation, with intense suffering, and final stiffening of the joint. While all this is going on, of course, the animal loses condition, and, if pregnant, may cast the lambs.

CHAPTER IX.

Sending for the veterinary surgeon—Imperfect messages—An important query—The conference—"Master has a horse took ill"—The mitigation of suffering—An appeal—Five suggestions—Laconic epistles—"Order is gain."

In the hurry and excitement consequent upon sudden illness among animals, especially when they are of great value, very unfit messengers are frequently selected and dispatched to summon the veterinary surgeon. A boy, or perhaps a woman, is called from the field; neither have seen the suffering animal, and probably do not know whether a horse, cow, a whole herd of cattle, or a flock of sheep are ailing. We have on many occasions had cause for lamenting the arrival of such messengers from farms many miles away, who have walked the whole distance, and the only information we could glean was that the messenger was exceedingly tired, and needed refreshment, and that he or she had been told to "hurry off for the doctor, and tell him to come directly." Sometimes, indeed not infrequently, when a verbal message has been sent, it has been either partially forgotten or so mutilated as to be entirely useless as information. Under these circumstances we can do nothing but go as soon as possible. We order the horse, and meanwhile look round the pharmacy—not to see what we can take, but really to learn, if possible, what we may safely and wisely leave behind.

In our dilemma we know not what to do. The chest in our conveyance already contains a miniature drug-shop, and adds considerably to the weight, as "Bees-wing" and "The Flying Dutchman" can testify. We, therefore, review its contents, pack in more doses, add extra "instruments of torture," pronouncing it full, and return to take another look round the pharmacy. Shortly our musings are terminated by the announcement, "The Dutchman's hat the door, sir;" and with a sigh of regret that we cannot transport the whole shop to the farm, nor prevail upon the farmers to set up one for the district, we step to the front door. Our assistant takes his seat, having securely placed the messenger behind. We also mount and drive off, the mind seriously occupied with conjectures as to the kind of case we are to meet with.

At length we arrive, and find half the village assembled. It is a congregation of the most sympathetic, good-hearted, earnest, honest fellows in the world, but wearing the most dolorous aspect of countenance possible. They surround the vehicle before we can descend; several seize the reins right and left, others lean upon the shafts, and some throw their arms across "the Dutchman." The principal speakers come nearer on each side—some rest a foot upon the step, and others, treating my vehicle as a manure cart, notwithstanding the new paint and varnish which has just cost me ten guineas, put their heavy, nailed boots on the nave and spokes of the wheels.

They are now fairly settled, and up to any amount of business, and the conference begins in real earnest.

"Good day, doctor. The woman there found you, I see."

"Oh yes, indeed. But why did you not send Tom Jones on one of the horses?"

"Didn't you meet un?"

"Meet un. Meet who?"

"Why, Tom Jones, of course."

"No, I did not. Why do you ask the question?"

"Becos we sent un off in horry to yer."

"Indeed. What for?"

"Why to tell you not to come, becos th' ould mare died about an hour agone."

"When was she taken ill?"

"About nine o'clock last night."

"And what was she like?"

"Oh, gripey, you know. She kept on all night, poor beast. She was horful bad, and we tried everything we could think of. At last, one said one thing, and another said another, and we agreed all of a suddin this morning about ten o'clock to send for you; and now you've come th' ould mare's dead after all. I can't think how Tom Jones could have missed you. We sent him on a horse, and telled him sure and make haste and stop you comin', and all that's no use. It's all ower true, sir, one bad job never happens without one or two more. You never know'd it, did yer?"

This is no imaginary case, dear reader, but a simple type of many which could be vouched for in the experience of every veterinary surgeon. It fails, however, to convey a tithe of the inconvenience, difficulty, and disappointment, to say nothing of the loss of time, to the practitioner, and the prolonged agony of the poor animal, which, under a different state of affairs, might have been saved. The illustration mainly serves to convey an idea of the serious consequences of delay; and we could add considerably to the testimony which proves the heavy mortality which might be avoided by seeking timely help for sick animals. It is far too common to allow a case to go far into the night, or even until the following morning, in the meanwhile often making use of the most unsuitable remedies, before medical help is sent for. If the messenger goes afoot, half the day is wasted, perhaps more, as the veterinary surgeon may have left for a long round in an opposite direction, and he only hears of the case on his arrival at home hours after.

The result of imperfect messages, delivered with tardiness, is painfully obvious in another way. We will explain our meaning by reference to another fact. The common order, "You're to come directly; master has a horse took ill," was delivered to us one afternoon about three o'clock. The messenger, a boy, had been sent direct from the field where he was at work, having to walk nine miles. In vain could we imagine what could be the matter, for the apparently urgent character of a message is not always a true representation of the case. The boy could divulge nothing more; we had no alternative but to go and see for ourselves. Upon arrival we found our patient was a *cow* instead of a horse; the poor creature was in difficult labour, but we had no proper instruments with us. Two hours or more were spent in the assistant going to fetch them; meanwhile we discovered that twins were present, and the owner, with others, had been busily pulling at the legs of both at the same time. After some severe work and great exhaustion the creature was delivered of dead calves, but she died some weeks after that long operation and needless delay, although she appeared at one time to have recovered from the effects of both.

The principles which should actuate all who have anything to do with animals should be the mitigation of suffering. It is purely the function of the veterinary surgeon, and he aims at that end by means with which he is specially qualified. He may be guilty of "doctoring for a living," and if he possesses no humanity, no sympathy for his patient under agonising pain from disease or accident, the man has selected a wrong calling. He should have been a mason or carpenter, so that he might cut, hew, and drive to the extent of his natural taste. When a man unites within himself an innate love of animals and a refined sense of sympathy for them under suffering, he not only makes a good doctor, but is excused for his selfishness in having selected a calling by which to gain his living. To such a man unwarrantable delay and imperfect messages are as paralysis to his hands and a perfect barrier against the good he wishes to execute; they vex his righteous soul from day to day as he sees the valuable lives thus sacrificed; while those who from personal interest alone should possess some portion at least of the sentiment look on almost with indifference.

On behalf of the many thousands of animals which, through absolute ignorance, selfishness, and positive want of feeling in far too many instances, are allowed to pass through needless and unreasonable agony, whose health and lives minister to the owner's wealth, we plead for more consideration in their bodily ailments; and with the view of strengthening the hands of the veterinary surgeon in the special means which he alone can exercise for alleviating animal suffering, we venture to make the following suggestions:—

All know the value of a nail; also its value as compared with that of a horse,—at least all think and say they do. Yet how often is the horse lost in the fruitless attempt at saving the cost of the nail. This is the main business with which many occupy the whole of their lives, and they earn the reputation of being careful, sensible, &c. How much less sense would there be in the man who, after ploughing and sowing his ground, refused to gather the crops because the cost of labour must be incurred? It is just probable we should declare him *non compos mentis*, and accordingly vote his removal to the nearest establishment where they take care of such people. Circumstances alter cases, we argue, and console ourselves with the false notion that it is no sin, nor yet a mark of insanity, to be indifferent about the suffering of animals, though that indifference may take the form of torturing them by all kinds of useless remedies. Animals have been produced only by cost and time, and both are as money to the proprietor, yet the suicidal policy we have satirised is still pursued in many places. There cannot, we think, be any doubt as to the similarity of the two instances, and we fear also there is little less than absolute cruelty exhibited in the method which characterizes much of the amateur medical treatment to which the patients are subjected.

We therefore respectfully advise:—

First. Whenever possible select a messenger who takes

an interest in the animal; at least one who has witnessed its sufferings, and is able to afford information which may prove useful to the veterinarian. If such a person is not at hand, or cannot be spared from important duties, and a stranger is selected, *send a written message.* Let us urge that on no account should a verbal message be intrusted to illiterate persons, as in all probability it would assume a totally different complexion before reaching its destination. If you are sending the first time to the veterinary surgeon, give full name and address also, and write as plainly as possible.

Second. Send *early*, or as soon as may be after the appearance of illness, so that the practitioner may see the case in its original state, and before it is rendered critical by dangerous complications. "A stitch in time saves nine."

Third. Send as much information as possible with regard to the symptoms exhibited by the animal, as well as any other prominent feature in connection with the occurrence. These need not make a long letter, nor take up much time, but may be exceedingly helpful to the practitioner in very urgent cases. The following laconic epistle was from a young farmer, possessed of good natural ability, but without any extent of education, who desired to profit by our instructions :—

"Ladywell Farm, 6 A.M.

"This is to inform you that we have just found the *red cow* standing fixed, won't move, and grunts awful. She seems quite bound in the bowels."

These facts, estimated in conjunction with the prevailing excessive drought, enabled us to take special remedies, and give the animal relief sooner than we might otherwise have done.

Another wrote as follows :—

"Five cows were found in the barn this morning, some eating the wheat, and others in great pain and blown. The men left the doors open when they went away last night."

A sheet of foolscap, covered on the four sides, could not have conveyed more intelligent language to the veterinary surgeon, who at once knew what serious cases he had before him. This message forms a strong contrast to one which we once received. It was simply—"Come at once and see the stirks, they are very ill." We went, and found upwards of thirty, some dead and others dying from eating the clippings of yew-trees. As will be expected, we had no suitable remedies with us, so had to return for them. Notwithstanding the delay, the success was greater than we had a right to expect.

Fourth. Always avoid giving medicines in the absence of sufficient knowledge of the nature of the malady and the required remedy. If you are sure the animal is suffering from colic, be equally sure you are giving suitable medicine.

Fifth. Whether any mistake has been made or not in the choice of remedies, do not hesitate to communicate fully everything in reference to them. Let it be remembered that our patient cannot tell us anything, and by your secrecy the suffering of the animal may be prolonged or aggravated; at the least it may not be relieved so quickly as desirable.

In the foregoing we have confined our attention to urgent and serious cases only. But it is not with them alone that needless delay and imperfect messages work irreparable mischief. By the observance of system and order we are convinced that the lives of thousands of animals might be saved which are now sacrificed on the "penny-wise system."

CHAPTER X.

Materia Medica—The action and uses of medicines; with their forms of combination—Alteratives—Anodynes—Antiseptics or antiputrescents—Antispasmodics—Astringents—Blisters—Caustics—Clysters—Enemas—Cordials—Demulcents—Diaphoretics—Digestives—Diuretics—Electuaries—Embrocations or liniments—Expectorants—Febrifuges—Fomentations—Lotions—Poultices—Tonics.

In order to produce the most beneficial and immediate effect upon the animal system, it is the usual practice to combine the remedies in certain order and proportions.

This can only be done by those who possess a knowledge of the action of drugs, as well as their chemical characters. If this is ignored, a compound may be made up of substances having strangely opposite qualities; it may, therefore, be useless as a remedy, but powerful as a poison.

The following are examples of the usual forms made use of, with the terms by which they are known. The practitioner is not confined to such a small list, but, by the application of his judgment, is able to enlarge it almost to any extent. We append also a brief outline of the action of the remedies, together with the doses in which they may be used. For lambs the dose must be reduced to one-half, or one-third, as age and size may dictate.

ALTERATIVES.

A great number of medicinal agents are included under this term, which is neither precise nor commendable. It is usually understood to apply to remedies which restore healthy action in the place of previously disordered function. Compounds of this kind are thus prepared:—

1. Epsom salts, powdered, 1 oz.; saltpetre, powdered, 1 drm.; ground ginger, 1 drm.; ground gentian, 1 drm. To be given as a drench.

2. Saltpetre, 2 drms.; sublimed sulphur, 2 drms.; ground gentian, 2 drms. Mix. A daily dose for mixing with the food.

3. Chlorate of potash, 2 drms.; sulphur, 2 drms.; linseed meal, 2 drms. Mix. Use as No. 2.

4. Fowler's solution of arsenic, 2 or 3 drms.; tincture of cardamoms, ½ oz. Mix, and add 6 oz. of linseed mucilage (see Demulcents), to make a drench.

ANODYNES.

Medicines known by this term are used to soothe and allay pain, quiet the nervous system, relieve spasm, &c. In conjunction with astringents, they are prescribed for diarrhœa.

1. Epsom salts, 1 or 2 oz.; extract of belladonna, ½ drm.; essence of peppermint, 3 or 4 teaspoonfuls; linseed mucilage, ½ pint. For simple colic.

2. Prepared chalk, 1 oz.; catechu, 4 drms.; ground ginger, 2 drms.; powdered opium, ½ drm. Mix, and add ½ pint of peppermint water. Dose, for diarrhœa, 2 or 3 tablespoonfuls, morning and night. See Astringents.

ANTISEPTICS OR ANTIPUTRESCENTS.

Substances known as antiseptics are those which arrest the tendency to decay, or putrefaction. They are required extensively in the treatment of diseases among animals, for the purpose of arresting the spread of infection, &c., which may arise from the unpropitious state of wounds, or discharges from inflamed mucous membranes. Under the term *disinfectants*, the same substances are more or less employed to purify the state of the air, drains, floors, woodwork, bedding, &c., &c., of buildings in which contagious diseases have existed among animals.

1. Any of the following acids, viz.:—Sulphuric, nitric, muriatic, or acetic, diluted with water in the proportion of 1 part to 100, form very useful antiseptics for wounds and chronic discharges.

2. Chlorine water, 2 oz. to a pint of cold soft water. To form a lotion.

3. Chloride of zinc, 3 grs.; distilled water, 1 oz. To form a lotion.

4. Lunar caustic, 5 grs.; distilled water, 1 oz. To form a lotion.

5. Pure carbolic acid, 4 oz.; pure glycerine, 4 oz. Mix and agitate, then add 28 oz. of pure olive oil, and agitate further to ensure perfect solution. Label "POISON."

6. Solution of sulphurous acid, applied as a lotion or by means of the spray-producer.

7. "Sanitas" oil, 1 oz.; pure glycerine, 4 oz.; water, 6 oz. To be used as a lotion or by means of the spray-producer.

8. "Sanitas" oil, 4 oz.; pure glycerine, 4 oz.; pure olive oil, 36 oz. Mix and agitate.

ANTISPASMODICS.

As the term obviously implies, antispasmodics are agents which relieve excessive muscular *spasm* or *cramp*. This condition, when affecting internal organs, &c., may depend upon the presence of irritants, and when they are removed the disease at once disappears.

1. Tincture of opium, ½ oz.; spirits of nitrous ether, ½ oz.; linseed mucilage, 6 oz. Mix for a drench.

2. Spirits of turpentine, 2 drms.; tincture of opium, 2 drms.; solution of aloes, 2 oz.; linseed mucilage, ½ pint. Mix, and give as a drench.

3. Spirits of nitrous ether, 1 oz.; powdered camphor, 1 drm. Mix and dissolve; then add—Tincture of opium, 2 drms.; powdered ginger, 2 drms.; and linseed mucilage, ½ pint. Mix, and give as a drench.

APERIENTS.

Aperients are mild or gentle purges. More powerful remedies are called *Cathartics;* and the weakest forms are known as *Laxatives.*

1. Linseed oil, 4 to 6 oz.

2. Linseed oil, 4 oz.; Croton oil, 5 drops.

3. Epsom salts, 4 oz.; ground ginger and gentian, of each ½ oz.; sublimed sulphur, ½ oz. Mix, and give in ½ pint of linseed mucilage. A moderate cathartic for large sheep.

4. Epsom salts, 4 oz.; ginger and gentian, as in No. 3; Croton oil, 5 drops. Mix, and give as No. 3. A strong cathartic for very large and strong sheep in severe constipation.

ASTRINGENTS.

Astringents contract the animal tissues with which they are brought into contact. They produce their effects in various ways: locally, by direct application; and generally, when administered by the mouth or introduced into the circulation.

EXTERNAL APPLICATIONS.

1. Goulard's extract of lead, 2 oz.; cold water, 1 pint.*

2. Sulphate of zinc, 1½ drms.; tincture of myrrh, 1 oz.; cold water, 1 pint. Dissolve the zinc in the water, then add the tincture.*

3. Sulphate of copper, 1 to 2 drms.; cold water, 1 pint. Dissolve.*

OINTMENTS.

4. Acetate of lead, 1 drm.; hog's lard,† 1 oz. Mix.
5. Sulphate of zinc, 1 drm.; hog's lard,† 1 oz. Mix.

ASTRINGENT POWDER.

6. Sulphate of zinc, 4 parts; oxide of zinc, 1 part; Armenian bole, 1 part. Mix. To be dusted over the parts daily.

INTERNAL REMEDIES.

7. Tincture of opium, 2 drms.; powdered catechu, 1 drm.; flour or starch, 1 oz. Mix rapidly, and add 6 oz. of tepid water. For a drench.

8. Powdered opium, 10 to 15 grs.; powered ginger, 1 drm.; powered alum, ½ drm.; strong tea, 6.oz. Mix for a drench.

BLISTERS.

Blisters are applications to the skin for the purpose of setting up irritation, in order to overcome or counteract already existing and deeper-seated inflammation. This is known as *counter-irritation*, or curing one disease by setting up another. The effect of a good blister is the production of large vesicles or bladders, when it is said to "rise." If the action is tardy, and blisters rise slowly or not at all, the sign is usually considered to be unfavourable.

LIQUID BLISTERS.

1. Olive oil, 1 pint; powdered Croton seeds, 1 oz.;

* To form a lotion. Label "*Poison.*"

† Free from salt.

powdered cantharides, 1 oz. Mix, and heat in a water bath for two hours. When cold, add spirits of turpentine, ½ pint, and allow the whole to stand twenty-four hours. Afterwards strain through muslin.

The following is extra strong:—

2. Bruised Croton seeds, 1 oz.; spirit of turpentine, 8 oz. Mix, and set aside for fourteen days, frequently agitating. Afterwards filter for use.

N.B.—Both these preparations must be used sparingly, so as to cause full absorption, as they are applied with friction. In other words, they must be well rubbed in, and not allowed to drain and saturate the wool below the shorn parts to which they are applied.

3. A good application for immediate use is the strong aqua ammoniæ. A few folds of rag, suitable to the size of the parts to be blistered, should be saturated with the solution, applied quickly, and covered by thick cloths and pressure.

OINTMENTS.

4. Powdered cantharides, 2 oz.; powered euphorbium, 1 oz.; oil of turpentine, 2 oz.; oil of origanum, 1 oz.; resin, 1 oz.; hog's lard, 16 oz. Place the cantharides, euphorbium, resin, and lard in a hot-water bath for eight hours. Remove, strain, set aside, and when getting cool add the turpentine and origanum. Agitate thoroughly, and allow it to set.

5. Tartar emetic, 1 oz.; hog's lard, 4 oz. Mix on a marble slab by means of a spatala. As this ointment is liable to spoil by keeping, it is necessary to make up only as much as required on each occasion.

N.B.—To insure the speedy action of blisters, the wool should be shorn as closely as possible immediately over the parts to be operated upon. The quantities stated in the foregoing prescriptions are sufficient to make many blisters; some discretion will, therefore, be required as to the amount to be used. The application should be attended with smart friction, and at the close some of the ointment should be left on, thinly coating the surface.

CAUSTICS.

Caustics are agents which exert a chemical action upon the living tissues, producing an effect equivalent to burning or decomposition. They are of two kinds: the *actual cautery*, or iron heated to redness; and the *potential cautery*, or mineral and chemical agents, as *caustic soda*, *potash*, and *lunar caustic*, or nitrate of silver. The heated iron is often the most useful, being employed for stimulating indolent wounds, abscising parts already sloughing, cutting off tumours, and arresting bleeding from an artery. The usual form is the firing or budding iron.

1. Caustic potash is conveniently sold in the form of pencils, having been fused and run into suitable moulds. A holder is required for using it. As a caustic, it is powerful and valuable; but, as it so quickly absorbs moisture, and becomes fluid, it is rather unmanageable, and likewise expensive.

2. Lunar caustic, or nitrate of silver, is by far the most manageable, and very effective. It is also sold in pencils, and requires a silver or platinum tube for use and conveyance.

3. Sulphate of copper, burnt alum, and verdigris in powder are used as dry caustics, for repressing fungoid growths and excessive granulations.

4. Muriate or butyr of antimony is a powerful caustic. The addition of water decomposes it.

5. Sulphuric, muriatic, nitric, and acetic acids are likewise valuable as powerful, and, with No. 4, may be applied by means of tow twisted upon the end of a stick.

6. Corrosive sublimate, 5 to 10 grs.; muriatic acid, ½ drm.; distilled water, 7½ fluid drms. An effective solution for injecting within fistulous sinuses.

OINTMENTS.

7. Verdigris, 1 oz.; hog's lard, 3 oz. Mix.
8. Sulphate of copper, 1 oz.; hog's lard, 4 oz. Mix.
9. Burnt alum, 1 oz.; hog's lard, 3 oz. Mix.

CLYSTERS OR ENEMAS.

Clysters are of two kinds, fluid and gaseous. The first are used for unloading the rectum and to convey nutrition to the system; the second are adopted to allay severe spasm.

FLUID ENEMAS.

1. Warm water, several quarts; soap (hard or soft), sufficient to make a strong solution by being rubbed down.

2. Substitute common salt for the soap.

Medicated enemas consist of some remedy added to water, gruel, or linseed mucilage.

3. Flour gruel, 1 pint; spirits of nitrous ether, ½ oz. Useful when the animal cannot take food.

4. Tincture of opium, 2 drms.; powdered catechu, 1 drm.; solution of starch, thickened by boiling, as used in the laundry, ½ pint.

5. Gaseous enema. Tobacco smoke, generated in a suitable apparatus, and passed up the rectum. When necessary, opium or assafœtida may be added.

CORDIALS.

Cordials are remedies having warm, stimulating, and tonic properties. They are useful as temporary stimulants, and for incorporating with other medicines to counteract their depressing effects. If the owner feels the necessity for such a preparation, he can make use of the following:—

1. Powdered carraway seeds, 2½ oz.; powdered ginger, 2½ oz.; oil of cloves, 60 drops; treacle, ½ lb.; warm linseed mucilage, 3 pints. Mix, and divide into 4 or 6 doses, as may be requisite.

DEMULCENTS.

Demulcents are substances which sheathe, soften, and soothe the parts with which they come in contact, and are particularly employed for lessening the irritation in mucous membranes, as those of the lungs, bowels, kidneys, and bladder.

1. Linseed mucilage. Linseed, 1 lb.; cold water, 1 gal. Mix, cover up, and set aside, frequently agitating. In 24 hours it is ready for use. Add warm water if required.

2. Linseed, 4 oz.; boiling water, 1 quart. Let the mixture simmer until a mucilaginous solution is obtained, then set aside to cool. This form is recommended when No. 1 is not kept constantly in use.

3. Gum arabic, finely powdered, 1 oz.; water, 1 pint. Agitate well, and administer in 2 or 3 doses.

DIAPHORETICS.

The power of certain medicines to act on the skin, producing a tendency to perspiration, is not so evident in the lower animals as in man, and among cattle and sheep perhaps the least of any. It is, however, possible to increase the circulation in the skin, and produce external warmth by means of stimulants, warm clothing, &c., and this course we advocate when necessary.

DIGESTIVES.

Digestives are occasionally used to promote suppuration as a means of securing the eventual healing of tardy wounds, ulcers, &c., and for dressing setons. The usual form is that of ointment.

1. Strong vinegar, 17 parts; honey, 14 parts; verdigris, 5 parts. Mix.

2. Verdigris, 1 oz.; Venice turpentine, 4 oz.; hog's lard, 8 oz.; resin, 1 oz. Melt the resin, then add the lard and turpentine. After these are thoroughly incorporated, add the verdigris, and stir frequently as the mixture cools.

3. Resin, 1 oz.; Venice turpentine, 2 oz.; hog's lard, 4 oz. Melt the whole and mix.

DIURETICS.

1. Diuretics are agents which promote a discharge of urine, and in this way reduce the watery parts of the blood. Such an effect is often desirable, as one of the means of promoting absorption of fluids which have been

effused in closed cavities, as in hydrothorax, and also in that form of cellular infiltration known as dropsy, &c.

DRENCHES.

1. Nitrate of potash, 2 drms.; gum arabic, 2 drms.; water, ¼ pint.

2. Nitrate of potash, 2 drms.; powdered digitalis, 8 or 10 grs.; linseed mucilage, ½ pint.

ELECTUARIES.

Electuaries are syrupy concoctions for conveying medicines to the mouth, where they are slowly dissolved by the saliva and swallowed. They are eminently suitable when the jaws cannot be separated, or when from sore throat, injuries to the fauces, &c., it is requisite to keep the parts in stillness.

1. Muriate of ammonia, 2 oz.; camphor, 1 oz.; gum kino, 1 oz. Each of the ingredients being reduced to fine powder and carefully mixed, add 1 lb. of treacle and thoroughly incorporate. Dose, one teaspoonful placed on the tongue twice or thrice a day.

2. Powdered catechu, 2 oz.; honey or treacle, 10 oz. Mix. Dose as No. 1.

EMBROCATIONS, OR LINIMENTS.

Embrocations, or liniments, are compounds prepared for *external use only*, being intended to produce a stimulating, sedative, or absorbing effect. Thus they are employed to stimulate the circulation and hasten nutrition in a locality previously weakened by disease; to reduce pain and inflammation; and, lastly, to disperse the enlargement left by previous disease.

STIMULATING.

1. Olive oil, 1 pint; liquor ammonia, 1 oz.; spirits of turpentine, 2 oz. Mix, and apply with smart friction.

SEDATIVE.

2. Extract of belladonna, 2 drms.; tincture of opium, 2 oz. Reduce the extract to an emulsion in a mortar by means of the tincture; afterwards add olive oil ½ pint. Apply with gentle friction.

3. Goulard's Extract, 1 oz.; tincture of opium, 2 oz.; olive oil, 1 pint. Mix and agitate to make a thick fluid. Use as No. 2.

STIMULATING AND SOOTHING.

4. Soap liniment (opodeldoc), 8 oz.; tincture of opium, 2 oz. Useful in later stages of acute and also chronic rheumatism as an application to the joints.

FOR DISPERSING ENLARGEMENTS.

5. Soap liniment (opodeldoc), 8 oz.; tincture of iodine, 6 oz.; tincture of opium, 2 oz. Apply daily with smart friction.

EXPECTORANTS.

Expectorants are employed to promote or excite a discharge from the mucous lining of the air passages. In disease of the lungs, bronchial tubes, &c., such action proves of the greatest benefit after the activity of the inflammation has been reduced. Remaining irritation, giving rise to a distressing cough, is frequently removed, and the animal gains rest and comfort thereby.

DRENCHES.

1. Aromatic spirits of ammonia, 2 drms.; extract of belladonna, 1 drm.; gum assafœtida, 1 oz.; linseed mucilage, 1 pint. Rub down the gum and extract with the spirits of ammonia and a small quantity of water to form an emulsion, then add the mucilage and agitate. Divide into four doses.

2. Spirits of nitrous ether, 2 oz.; oxymel squills, 4 drms.; extract of belladonna, 1 drm.; linseed mucilage, 1 pint; water, a sufficiency. Mix as directed for No. 1. Divide into four doses.

FEBRIFUGES.

Febrifuges, or fever medicines, are made use of to reduce that condition known as "fever," which accompanies all acute diseases, and forms a prominent indication of their character. The action of the heart is reduced, all important secretions are augmented, and the blood is deprived of the materials which favour the process of inflammation.

DRENCHES.

1. Nitrate of potash, 1 oz.; camphor, powdered 2 drms.; digitalis, powdered, ½ drm. Mix, and divide into four doses, each to be given in 6 or 8 oz. of linseed mucilage.

2. Solution of the acetate of ammonia, 4 oz.; tincture of belladonna, 1 oz.; linseed mucilage, 1 pint. Mix, and divide into four doses.

FOMENTATIONS.

In order to obtain the needful benefit by the employment of hot water, in the restoration of action in parts injured by disease or accident, the following directions should be scrupulously followed:—

The temperature (about 118° Fahr., and not higher than 120° Fahr.) should be constantly maintained throughout; a plentiful and constant supply of water being provided.

The use of fomentations should be prolonged, in some cases as many as four or six hours being occupied.

The parts should be covered from the first by means of a rug, piece of flannel, blanket, &c., several times folded, which, at the close, is to be replaced by dry portions. Subsequent cooling, especially after short and careless fomentation, often proves as hurtful as the disease.

All things being favourable, and the animal suitably placed, a large pail or open tub is brought as near as possible and filled with hot water, the temperature denoted by a thermometer partially immersed. The diseased parts being covered, as already advised, by the materials wrung out of the hot water, the operations then commence by keeping up a constant stream poured on at the highest point, by means of a can, &c., holding about a pint or more. Other assistants should be respectively deputed to keep up a fire beneath a copper for the supply of hot water, and to carry it when the distance is great. The water may be economised if the stream descending from the coverings is caused to flow into the open tub beneath; but, under all circumstances, the supply of hot water must be constant in order to keep up the temperature.

It is sometimes necessary to use *medicated fomentations*. These are merely hot water containing solutions of certain medicines, infusions, &c., used with the view of producing an additional effect upon the diseased parts. Thus, in severe pain and inflammation, rheumatism of a joint or limb, laudanum or poppy heads or seeds are added to produce a soothing effect. When there are foul wounds and ichorous discharges, as rat-tails or psoriasis, foot-rot, &c. (see chapter xxvi.), 2 oz. each of soft soap and spirits of turpentine will aid the temperature in stimulating the parts to healthy action, besides cleansing them effectually. Carbolic acid (Antiseptics No. 6, 7, or 8) may also be added. In order to soften and break down incrustations of the skin, due to previous long-standing inflammation, &c., and admit of the action of other remedies more immediately, such as in the case of rat-tails, referred to above, a mixture of glycerine and potash will accomplish speedy effects. To a pail of water, 2 oz. of glycerine and 1 oz. of carbonate of potash will be ample.

LOTIONS.

Lotions are preparations made by the solution of certain substances in water or spirits, for the purpose of cooling the parts to which they are applied. In this way inflammation is reduced and tone imparted to weak joints, tendons, &c.

1. Tincture of arnica, 1 oz.; spirits of wine, 7 oz. Apply with moderate friction to insure absorption.

2. Goulard's Extract, 4 oz.; dilute acetic acid, 2 oz.; distilled water, 1 quart.

3. Solution of the acetate of ammonia, 4 oz.; spirits of wine, 4 oz.; water, 1 pint.

Nos. 2 and 3 may be applied, by means of a sponge or rag, repeatedly during the day, or the parts may be covered by a bandage, &c., kept constantly wet with them.

HEALING LOTION FOR WOUNDS.

4. Sulphate of zinc, ¾ oz.; sugar of lead, 1 oz.; tincture of myrrh, 2 oz.; soft water, 1 quart. Mix, and shake well before using. The lotion should be *dashed* upon the surface of the wound direct from the bottle.

POULTICES.

The action of poultices is frequently an effective auxiliary in the treatment of disease. The object of their employment is twofold:—

To apply continued heat and moisture in order to soften and cleanse parts, and promote circulation and suppuration as thus conducive to the healing process;

To maintain a low temperature as cold as may be required by the nature of the disease.

It is not generally known that by a careless and improper use of poultices a great deal of harm follows. Thus, after warmth has been supplied for some time, the poultice is removed to make way for another; but meanwhile the parts are unprotected, and as cold constringes the vessels, the previous effect of heat—that of relaxation—has been destroyed. Again, when the application of cold is transient and spasmodic, reaction takes place. Parts in which a constringent and tonic action is called for suffer increased relaxation, and curative progress is greatly delayed.

In order to derive the proper benefit from a poultice, the application should be continuous. When removal is called for, to test the progress of curative action, or when a change is needful, a fresh poultice should be prepared, so as to take the place of the first immediately on its liberation from the parts. This is the only way to maintain the desirable temperature.

With regard to the composition of poultices, there is ample choice. They should always be *cleanly*. This is not generally the case, for we have, on many occasions, been subjected to the most trying ordeals by the use of such disgusting applications as cow-dung, and even human ordure to the wounded limbs, &c., of animals. The excuse is, "they clean the parts wonderfully." This may apparently be so, as a result of the saline constituents which excrement contains; but we need not resort to such filthy concoctions for that purpose. We have only to refer our readers to the effects which are observed in the feet of sheep constantly standing in poultices of this kind—the filth of the folds, pastures, straw-yards, &c., which produce one of the greatest banes of the sheep-farmer, viz. foot-rot—and the truth will be obvious.

The application of filth to a wound is the very thing which experience teaches us to avoid. It may, nay frequently does, produce serious blood-poisoning, the noxious elements of putrefaction being absorbed into the system. This is constantly going on in other ways, to the sad destruction of the farmer's profits, meaning so much loss to the nation, as well as dependence upon the foreign producer. We allude to the many blood diseases arising from poison germs given off by decaying matter, chiefly animal, though vegetable organisms are not without share in the blame. Seeing this, are we wise in courting disease by imitating the process of infection in the use of such filthy compounds as excrement of any kind for poultices?

We have said that the choice of materials for poultices is by no means limited. *Bran* forms a very useful agent, and if we require to make it a little more plastic, a handful of linseed meal or wheaten flour will render this effectual. Sometimes poultices are made entirely of *linseed meal*, as they retain heat longer. The objections are that they are heavy, and inclined to become hard in drying. The latter is removed by adding glycerine, linseed, or other oil, before the hot water is put in. *Ground linseed* is far preferable to linseed meal, the former containing the natural oil, the latter being linseed cake ground, and, therefore, contains only an appreciable quantity.

One of the most useful and effective agents, possessing all the desirable qualities of a poultice and fomentation combined, is *Spongio piline*, which consists of a sheet of impervious or waterproof material, thickly coated on one side with wool. If this is dipped in hot water it may be bound almost upon any part of the body, the patient suffering no discomfort from weight or rapid cooling. We have sometimes directed a stream of hot water (118° F.) beneath this material, with the most satisfactory results. *Spongio piline* is made in various thicknesses. For use among animals it should not be less than three-quarters of an inch, the superficial area being larger than the diseased parts, and the wool side placed in contact with them.

Medicated poultices are frequently required. These are chiefly prescribed by the veterinary surgeon, and consist of the simple kind we have named, to which some active medicinal agent has been added. The common forms are those in which Digestives (page 91) are thoroughly incorporated by mixing with hot water, for the purpose of promoting suppuration in indolent wounds of the feet, &c. Sometimes Disinfectants, No. 2, 6, 7, or 8, page 85, are added to destroy foul smells which emanate from such wounds, and prevent blood-poisoning. Other varieties will be prescribed by the veterinary attendant. See "Fomentations," page 94.

TONICS.

Tonics are those agents which promote strength or tone and vigour of constitution. They are always useful in recovery from disease, but require care in their prescription.

TONIC DRENCH.

1. Saccharated carbonate of iron, 1 oz; powdered gentian and ginger, of each 2 oz.; linseed mucilage, 1 pint. Divide into 3 or 4 doses.

TONIC POWDERS.

2. Saccharated carbonate of iron, 1 oz.; gentian, 1 oz.; powdered locust bean, 1 oz. Mix. Divide into 4 powders, one to be given in the manger food morning and night.

VEGETABLE TONICS.

3. Powdered gentian and colombo, of each 1 oz.; cinchona bark, 2 drms.; or quinine, 20 grs.; ground ginger, 1 oz. Divide into 4 doses, which may be given in the food, as No. 2, or as a drench. When strong tonics are required, these may be added to either of the above.

CHAPTER XI.

Blood Diseases arising from deranged or inordinate functions—Plethora—Anæmia—Hunger rot—Rheumatism—Uræmia—Tubercular consumption—Pining—Apnœa in sheep and lambs—Goitre, or "Derbyshire neck"—Rickets, or softening of bone.

THE maladies we now proceed to notice owe their origin to causes which derange the functions of nutrition and depuration, and, more or less, also give rise to an inordinate action in others. Among the causes we must also include hereditary taint, though primarily it may be due to influences which we now commonly regard as effects.

Plethora.

Fulness of Blood.—This is shown to consist of an excess of rich food materials in the blood, giving rise to rapid growth, improvement, and "blooming condition." These appearances are so unusually and quickly developed as to excite attention and surprise, and if no change takes place in the system of feeding fatal disease is engendered. At this stage the system is liable to contract blood-poisoning from the excess of its own materials; and escaping this, equally serious congestions of important organs may ensue. Such signs should warn the proprietor against premature laying on of flesh, particularly when sheep are confined to the fold or to small and rich pastures. Pregnant ewes should be carefully scrutinised, and the least signs of plethora met by prompt and effective measures.

Food of a less stimulating kind should be substituted in all cases. The over-luxuriant pasture must be changed for one in which, during greater part of the time, the sheep must work for their living. Those confined to folds and small enclosures will require immediate attention. An aperient, No. 3 or 4, page 86, should be given to each animal, and the veterinarian consulted as to the advisability of a course of neutral salts, &c.

The tendency of plethora to develop serious forms of disease is not so generally understood and recognised as it might be. This evil is probably more marked in the sheep than in any other animal. The duration of life is short, and in order to fit it for the condition in which it is required as flesh-food, the system is forced and hastened through successive stages of artificial development, by equally artificial means, until the process ends in the application of the butcher's knife. In order to produce sound nutritious meat, exercise is as needful as food, and the withholding of it tells as speedily as the want of aliment. In reference to this subject we quote from Professor Robertson:—*

"There are many of our most valuable ewes on our best lands, if not kept the whole time from conception until parturition on a full and generous diet, are at least during a greater portion of this time supplied with a full quantity of turnips, first on their pastures, and latterly

* "Hints to Stockowners." Blackwood & Sons, 1870.

on the land where these are grown. As it is well known that our Leicester ewes are voracious feeders, and also that exercise is absolutely necessary to the enjoyment of health in all pregnant animals, breeders have endeavoured to counteract the evil effects of this propensity on the one hand, and on the other to secure as far as possible the advantages resulting from the laws of health, by removing their ewes from the turnip brake during the day, and returning them at sunset. Could we insure that during the daytime, while on their pastures, the stock were kept moving about as they are wont to do when ordinarily grazing there, it were all very well, and the object would be attained. In most of these cases, however, the amount of exercise taken by the ewes is insignificant. The inducements for them to roam in search of food over the pasture lands is small. There is little calculated to please their palates after they have been folded on turnips, and they very soon seem to acquire the knowledge that in a few hours they will be taken to the roots left in the morning. The consequence of this is, that when driven from the turnips they quietly rest themselves by the side of the fold until the hour returns when they are again admitted, when, with appetites quickened by their self-imposed fast, they determinedly gorge themselves with enough upon which to rest and ruminate until the period again occurs for the repetition of the act. In this way the principle sought to be obtained by this division of time between pasture and turnips is, to a great extent, if not altogether, lost; and another train of consequences very undesirable is the result of the repeated gorgings to which the animals become addicted."

Anæmia.

Deficiency of Blood.—The bloodless state is due to various conditions which drain the circulation of its normal nutritious elements. Frequent blood-lettings, excessive use of debilitating medicines, deficient food, or a supply of that which contains little or no nourishment, wasting from continued disease or excessive yield of milk, are among the causes. Breeding ewes are common victims where gross mismanagement exists, if they are allowed through want of food to lose condition at the time they become pregnant. The demands now made upon them are excessive, and no amount of good food will restore the necessary condition, especially when they are drained by the subsequent lactation. Weakness and emaciation proceed rapidly; the wasted muscles are flabby, pulse weak and small, heart sounds audible, appetite lost; the abdomen swells, noises are heard within, and flatus is constantly passed; the action of the bowels is irregular, constipation and diarrhœa alternating with each other. The mucous membranes are pale, the eye exhibiting the blanched appearance so well observed in "liver rot," that, in contradistinction, anæmia, as seen in ewes, has in some districts been termed the "hunger rot." Dropsical swellings appear; and when this takes place between and beneath the jaws, the animal is said to be "poked" or "chockered." Life is usually terminated by diarrhœa or dysentery.

Treatment.—Remove the cause. Supply moderate quantities of good food, fresh air, and water. Tonics are called for. If diarrhœa is present, treat as for that disease (chapter xiv.).

Rheumatism.

Cold Felon, Chine Felon, and Joint Felon.—This is a disease which depends upon a peculiar state of the blood; it is charged with certain elements inimical to its constitution, and which are not removed, probably owing to an arrest of function. A perverted assimilation may be charged with its production, that in its turn being induced by errors of diet, and neglect of precautions both sanitary and hygienic. It is true that no disease exhibits greater proofs of its hereditary nature; but it is equally true that successive generations of animals come and go without its appearance, and suddenly, when most unlooked for, it becomes a bane to the sheep-breeder—the old sheep are crippled and the lambs are dropped without the powers of locomotion. One of the greatest causes of rheumatism is the too sudden change from the heat and confinement of close and confined folds or paddocks to cold, and perhaps wet and undrained, land. This acts, also, with double force on those already affected. In the usual form the joints are the seat of the affection in adult animals, and chronic stages are marked by complications such as disease of the heart and its coverings. In lambs, as will be shown in chapter xvi., awkward complications arise, and shortly prove fatal. The disposition to fly from one joint to another is characteristic, and thus the sufferings of the animal are rendered continuous, protracted, and acute. Ultimate stiffening is usual if the creature survives, but in many cases emaciation proves the incentive to humane slaughter. This is one of the many diseases which breeders would do well to study, in company with the veterinary surgeon, with the view to prevention.

Treatment.—Strong and vigorous animals only can endure the action of needful remedies. Aperients Nos. 1 to 4, page 86. To the joints, in the acute or early stages, Embrocations Nos. 2 or 3, page 92; afterwards, Nos. 4 and 5. Internally, during febrile disturbance, after the bowels have been moved, Febrifuges Nos. 1 and 2, page 94, given two or three times daily, as required.

Uræmia.

Blood Poisoning, due to the retention of those elements which should pass out with the urine. In fully developed cases the skin exhales an intolerable smell of urine. The mouth, breath, and fæces are likewise offensive; and if the animal does not obtain relief, he becomes dull, listless, and at last insensible, from which he rarely recovers.

The cause should be removed promptly to insure success. See chapter xxv. As the causes are sometimes obscure, the advice of a veterinary surgeon is needful, not only to cure the affected animal, but to advise as to the prevention of the spread to others.

Tubercular Consumption.

Tuberculosis, Consumption, Phthisis, Consumption of the Lungs.—This form of tuberculosis is not so common in sheep as in cattle. Animals of spare and attenuated form, having long legs, a narrow chest, &c., &c., are the usual victims, although hereditary taint may slumber in animals of a better formation for generations, and on their successors being subjected to questionable practices and violation of the rules of hygiene, the disease at once manifests itself. The causes are imperfect and innutritious food at the time when there are unusual demands on the system, as in conception, lactation, &c., which give rise to perverted assimilation and other deranged functions. By these means the system is loaded with materials incapable of supplying the exorbitant waste of the system, and they take the form of deposits or tubercles in various parts, as the lungs, and beneath the lining membrane of the chest, as well as in the lymphatic glands, &c. They are to be met with in all stages of growth, varying from the simplest form of redness and congestion to the hard grey tubercular or calcareous and soft cheesy-looking structure, which denotes softening, and preparation for abscess, &c.

The more common form is known as "Pining," and is simply the location of the disease in the glands of the intestines. Another form is indicated by tumours in the throat and neck, which sometimes discharge an offensive matter, and heal tardily. Lambs exhibit the disease at birth in some localities, apparently associated with rheumatism, the bones of the joints being affected. This will be referred to in chapter xvi.

Ewes affected with tuberculosis are weak, and at length emaciated; the function of respiration is embarrassed, and the creature dies of inanition. The "piner" is "pot-bellied," lanky, weak, emaciated, and, with the above, exhibits a tendency to local dropsy. The disease is said to prevail upon soils over the igneous rocks, but disappears when the animals are removed to the sandstone formations. Diseased animals should not be used as human food—the flesh is only fit for burning or burial. The milk of ewes affected is but another method of transmitting the disease to the offspring, though it is probable that it may be diseased at birth.

Treatment must be preventive.

Apnœa.

Difficulty of Breathing.—This is in reality a blood disease, and it is high time an appellation consonant with its nature be given to it. The particular form indicated arises in sheep and lambs: in the first it is due to the matting down of the fleece by means of fats and ungents and the subsequent accumulation of dirt, &c.; in lambs it arises from wearing the skin of another, put on for the purpose of deceiving the ewe and inducing her to take to a motherless lamb in place of one she has lost.

In both cases the action is nearly alike. The first effect is, the ointments cause the fleece to lie close to the skin, and become quite impervious to its exhalations, The skin of the dead lamb after a time fits so closely that the two may be said to resemble a macintosh, which when worn too long by the human subject proves similarly injurious. As the skin materially assists in the purification of the blood by giving off perspiration, sensible and insensible, it will be at once understood that by covering it with an impervious coat will, of course, stop that most important function. The blood is, therefore, loaded with hurtful materials which prove poisonous, and in addition the system is further poisoned by absorbing the active principle contained in the ointment, or the poison of putrefaction arising from the skin of the dead lamb.

Treatment.—Wash the ointment from the fleece with warm water having potash dissolved in it, or in severe cases shear the sheep at once, and then wash and apply friction to the skin by means of coarse towels or a soft brush. Remove the skin from the lamb wearing it, wash the body in tepid or warm water, followed by friction, and in both cases give nitrous ether as a stimulant when depression is great. Put the animals into comfortable and airy quarters, and avoid cold by means of light warm clothing.

Goitre, or "Derbyshire Neck."

This is a disease purely dependent upon geological formation of certain districts. It is characterized by a swelling of the thyroid body, which by gradual extension sometimes follows the median line of the neck from the throat, where it is first seen, to the front of the breast. It is a disease common to man and the lower animals, and prevails in districts lying on the magnesian limestone rocks, and in those contiguous if supplied with the same water, although the soil is altogether different. In this country it is mostly seen in Derbyshire; hence the title given above. The swelling is at first soft, or not unlike dough under pressure; but as the disease and enlargement proceed it becomes hard and tense or unyielding, and when divided by the knife is found to have a cartilaginous nature, with an admixture of calcareous or gritty particles. The disease is hereditary. Ewes transmit it to their progeny, which are usually born diseased, in many cases already dying or already dead. From this cause sheep-breeding in limestone districts is often rendered most unprofitable.

Treatment.—Remove the flock, especially breeding ewes, from the limestone rocks to soil on clay or sand, &c., if possible, but secure a supply of *soft* river or rain water under all circumstances. Native water may be neutralised and rendered soft by adding solution of carbonate of soda or potash, as long as a precipitate is thrown down. When the flock cannot be removed they should be restricted to rain water as a drink; breeding ewes are not otherwise safe. The crops of good sand, clay, or other soils should be supplied on the ground under these circumstances, and good roots must not be forgotten.

The best application to the swelling is iodine ointment. Iodine should be given internally when the stomach is empty, and weak solutions may be injected within the substance of the swelling. If the preventive measures are not successful, we have little hope of doing much good by medical treatment to so many diseased animals under present circumstances.

Rickets, Softening of Bone.

When breeding ewes are confined to pastures on porous or sandy and poor soils, they are liable to contract a condition of anæmia in proportion as the scanty produce is allowed to suffice for food. The results are probably traceable no farther in the mother than simply the low condition, but when the lambs are born they are too weak to stand, the bones bend and twist in unsightly forms, dislocation of joints takes place, the creature has not much desire for food, and the ewe's milk fails to nourish effectually. Under these circumstances it is not uncommon that the whole fall of lambs is lost. In other instances the lambs are seized at a later period of life, when the same causes which established the predisposition during pregnancy operate upon their immature constitutions, and at once develop the malady.

Treatment.—This must be essentially that recommended for chronic indigestion, chapter xiv. See also Anæmia, chapter xi. Remove the ewes to fertile soils, and allow *judicious* supplies of grain or meal, so as to improve the milk, or resort to hand-feeding with good cow's milk, and give daily doses of cod-liver oil, pepsine wine, &c. The disease is to be prevented by due care of the breeding ewes. Land on which the disease is known to prevail should be improved by top dressings containing phosphates at suitable seasons, such as bones, superphosphate mixed with clay (raw or calcined), salt, &c.

CHAPTER XII.

Diseases of the Circulatory System—Anæmic palpitation—Rupture of the heart—Cyanosis—Inflammation of the heart—Foreign bodies in the heart—Pericarditis—Endocarditis—Enlargement of the heart—Dilatation—Fatty degeneration—Displacement of the heart—Embolism.

OUR notice of the diseases of the circulatory organs must necessarily be brief. The short term of life which is allotted to the sheep relieves that animal from many causes which give rise to special diseases of organs common to other creatures; in addition, the manner of forcing a system on to the marketable stage probably tends to lessen disease in one direction by creating it in another—at least, this seems not at all improbable in reference to the prevalence of the affections we are about to consider.

Anæmic Palpitation.

The Thumps.—We have already referred to the audible heart sounds under Anæmia, page 101. They are by no means to be always regarded as due to structural disease of the heart. The constant "thumping against the walls of the chest," as this phenomenon has been described, results from the watery state of the blood, which renders it an inelastic and unyielding fluid, which is the reverse of health. At each contraction of the heart the discharge of blood is forced into contact with the non-elastic column already within the relaxed vessels. The first is rising, the second descending. Sound is thus developed, and vibration conveyed throughout contiguous organs. Other signs are, paleness of the membranes, with advance of local dropsies in the early stages, and before emaciation sets in. The palpitations, which are intermittent, depend upon some sudden excitement, are regular in the beats, and occasion jerking of the abdomen. Abnormal heart sounds are absent.

If structural disease of the heart is present, the disease is developed slowly, unassociated with breeding, poverty, and bad management. There are, in addition, abnormal heart sounds, red mucous membranes, and dropsy of the limbs. This form we can hardly expect to witness in sheep.

Treatment as for anæmia.

Rupture of the Heart.

Rare as this untoward accident is among horned stock, it is still more rare among sheep. We believe it has taken place in aged sheep during combat, but further evidence is wanting. Death was, of course, immediate.

Cyanosis.

The Blue Disease.—This is due to an irremediable state of the heart, which admits of the mixing of the arterial and venous blood after birth. The separation between the upper cavities of the heart is not complete as it should be, or shortly after birth the communication may be re-opened, owing to obstruction in the circulation, such as may arise from tuberculosis, congestion, &c. The result is an admixture of the pure with the impure blood, and as it circulates through the system a blue colour is communicated to the tissues, followed by general coldness, prostration, and finally death. Common humanity suggests slaughter as a means of relieving the animal from suffering.

Inflammation of the Heart.

Carditis.—Inflammation of the substance of the heart is not known to exist to any large extent. If the entire organ was seized its functions would be so seriously interfered with as to occasion instant death. The outer surface is inflamed in connection with pericarditis, and the inner surface as endocarditis, the morbid action being limited to the membranes from which these diseases take their name. If the substance is affected, the disease is localised, and an abscess probably results.

Foreign Bodies in the Heart.

Sheep, as well as cattle, are liable to peculiar taste with respect to foreign substances. They will swallow very strange things, but, removed as they are from proximity to dwellings, the opportunities are not frequently put in their way. When sharp-pointed bodies, as needles, nails, skewers, &c., enter the stomach, they are carried about during the motions of the injesta and the organ itself until they are caused to enter the walls. The direction being usually forward, they are pressed onwards and finally reach the heart, where they are lodged, producing inflammation, abscess, and death. Although signs are indicative of heart disease, these are considerably modified by complications, and it is often only after death that a correct opinion can be formed.

Pericarditis.

Inflammation of the Heart Bag.—This disease may arise in consequence of puncture by foreign bodies, as explained in the foregoing paragraph, or it may have no apparent connection with these or other causes. As a rule it is due to the existence of rheumatism, with which it is complicated. The natural processes of circulation and respiration are very seriously impeded, sympathetic fever runs high, the pulse is rapid and wiry, the nostrils are dilated, secretions diminished, and constipation is present; the head is held low, the ears droop, and the sufferer will neither move nor lie down, if he can avoid doing either, on account of the extreme pain he endures; friction sounds are evident in the region of the heart, which only the veterinarian can interpret, and the jugular vein exhibits the signs of pulsation caused by the obstruction in the heart. As the disease progresses the friction sounds are no longer heard—they are suppressed by effusion of water, which swells the bag, oppresses the heart, and eventually suffocates the patient in the course of four or five days. Less severe cases go over this time and usually recover.

Treatment.—The propriety of bleeding is much questioned, and greater reliance is placed on powerful internal remedies. Febrifuges No. 1 or 2, on page 94, as required. Blisters to the sides and front of the chest Nos. 1 to 5, pages 87, 88. Aperients during constipation Nos. 1 to 4, page 86. Diuretics No. 1 or 2, on page 92. Warm coverings, fresh air, nitrated water, light food. When the pulse subsides, and other indications of the arrest of the disease are present, tonics may be carefully introduced, either as powders in the food or as a drench.

Endocarditis.

Inflammation of the membrane lining the cavities of the heart does not exist, except very rarely, as an independent affection. Like pericarditis, it is usually associated with rheumatic forms of disease. There are all the intense signs of constitutional disturbance and interference with the heart's action. The pulsations are irregular and intermittent, the pauses between the beats being of variable duration. The smallness of the pulse is characteristic, as also the bellows-sound of the heart. If both the ventricles are affected the venous pulse will be present, and the difficulty of breathing very great. Suffocation usually carries off the animal.

The *treatment*, which is essentially that required for rheumatism, should be prompt and vigorous.

Enlargement of the Heart

Is not so common as in the ox, except where the sheep are overfed and the cramming system enforced to a great extent. It is due to some obstruction in the circulation such as disease of the brain, but more generally may be traced to the accumulation of fat upon and around important organs. Although the enlargement of the substance of the heart may take place throughout the organ, as a rule one side only is affected.

Dilatation of the Heart,

Chiefly confined to one side, also arises from the same causes as give rise to enlargement. It produces a total lack of power, appetite, and spirit; difficult breathing, weakness, palpitations, coldness of the limbs, &c. In this and the foregoing affection the animal must be kept very quiet, and fatted for the butcher with careful measures. The forcing system must be abandoned.

Fatty Degeneration of the Heart.

In this disease the substance of the muscular tissue is replaced greatly by fat, and in addition there may be other derangements, such as dilatation, &c., as well as degeneration of the general muscular system. Pampered animals of the various breeds are liable to the disease, but it is said that the Leicesters and Southdowns are the most common subjects.

Displacement of the Heart

Is not an infrequent occurrence among lambs at birth. The situations to which the organ is removed are as variable as they are frequent. Sometimes it is altogether outside the chest.

Embolism.

In consequence of vital changes taking place in the constitution of the blood, and, possibly, the admission of particles altogether foreign to it, occasionally, also, from causes which produce inflammation or degeneration of the walls of the vessels or their lining membrane, fibrine accumulates at the precise spot, and eventually it may stop the entire circulation. In endocarditis, clots or shreds of fibrine may be carried into the vessels and plug up a smaller branch. The result is loss of nutrition, wasting, coldness and paralysis of the part or whole limb below the obstruction. The disease is incurable.

CHAPTER XIII.

Contagious or Epizoötic diseases of Sheep—Epizoötic aphtha, or foot-and-mouth disease—Small-pox—Measles, or rubeola—Means of prevention—Ventilation—Disinfection—Fumigation—Disposal of manure.

THIS class of diseases, fortunately for the sheep-breeder, is not so large as the analogous one in reference to cattle. Notwithstanding that so much has been asseverated, rinderpest and contagious pleuro-pneumonia of horned stock have not been witnessed in sheep. The maladies we have to consider are thus reduced to those named above.

Epizootic Aphtha,

Or the *Foot-and-Mouth Disease.*—This is a highly contagious malady, communicable to several other species of animals, and by them also to sheep. It belongs to the class of eruptive fevers, dependent upon the introduction of an animal poison, and occurring, as a rule, but once in the lifetime of the sufferer. It is of foreign origin, and, but for the continued importation of fresh germs along with cattle and sheep, the extremes of temperature common to our variable climate would alone destroy the poison. As it is, we suffer periodically from visitations thus introduced, which cost the nation many thousands annually.

Like all true contagious maladies, epizoötic aphtha observes a period of incubation, which may not extend beyond twenty-four hours, but, as a rule, it occupies three or four days. We have had our attention drawn to animals on a farm when the disease has been raging in the surrounding neighbourhood, and, during three days before the earliest outward signs have appeared, an elevation of temperature, as exhibited by a thermometer placed in the rectum, was especially noticed. The first indication beyond this is a shivering fit, or a succession, followed by a staring coat, slight husky cough, and increased frequency of pulse, with other signs of ordinary febrile disturbance. The affected animal separates itself from the flock, appears dull, and disinclined to eat. Febrile symptoms grow intense, a flow of saliva and mucus flows from the lips when separated, and vesicles may be observed upon the tongue and gums, as well as upon the sides of the mouth. The creature becomes uneasy, and moves the jaws and feet, and shortly the flow of saliva is increased; it is also thicker, and discoloured by an admixture of blood and mucus, which hangs like so many ropes as the creature champs or smacks the lips. The vesicles have now burst, and occasion much pain and inability to eat. Sometimes the feet are the first to exhibit the blisters, when the coronet and other tissues of the foot are inflamed, the process also extending more or less towards the fetlocks. Lameness is, therefore, present, and, in addition, the animal kicks out or shakes the hind feet as it walks; occasionally the pain is so great as to cause the sheep to lie almost continually, especially after the vesicles burst and expose so much raw surface. In ordinary cases, the severity of the signs begins to subside about the fifth day, the raw surfaces are covered by epithelium, and by the fifteenth day the sheep is convalescent.

Severe cases are remarkable for tardiness in healing powers, and consequently, also, prolongation of suffering. Ewes suckling their lambs suffer acutely, the udder and teats partaking of the disease so as to destroy entirely the secretion of milk. The lamb takes the disease, and usually dies. Pregnant ewes abort, and thus the loss is manifold. Sloughing of tissues now prominently marks the disease. The nasal passages, tongue, and gums are a mass of ragged, raw, and ulcerated parts; the skin and hoofs have left the fetlocks and feet, and the creature crawls on bended knees; occasionally a whole phalanx of bones are removed, and the stump is presented to the ground. Abscesses form in the udder, followed by sloughing, and even mortification. In addition to these local states, the lungs partake of the congestion, the breath becomes fœtid, the eyes are bloodshot, and agony is extreme. Emaciation has set in, the blood is no longer able to nourish the tissues, and the heartbeats are loud and thumping; abscesses form over the body, and leave large ugly wounds; the bowels take on the disease, and colic, with diarrhœa, ends the life of the sufferer, after varying periods of illness ranging from one to three or five weeks.

Treatment.—No greater mistake can be made than to suppose this or any other contagious disease can be cured. No sane man, knowing anything about them, will make such an assertion, unless he is a rogue, and desires to live at the expense of those whose animals are suffering. But the disease admits of alleviation; its effects may be greatly mitigated by proper treatment; and this is all that honest practitioners attempt to do. At the present moment there is no remedy known that has the power of destroying the blood poison of contagious diseases; unless we know of such an agent, it is useless to speak of a cure for them. Such is their nature that acute observers have proved, in consecutive instances, that, whether treated by medicines or not, the malady still goes through its prescribed or usual course, and terminates no sooner than when by the sole efforts of nature the poison is eliminated from the system. By treatment only one result is gained; but it is worth gaining. This, as we have already stated, is an alleviation of the sufferings of the miserable creature, as well as supporting it under the depressing effects. By supporting the system we mitigate the severity of the disease, so that greater progress is made when the crisis is passed.

Palliative Triatment.—Remove the animals to a dry and roomy fold or yard for shelter, the ground being covered with short clean straw. Divide it through the centre by means of hurdles, those at the end being left loose so as to pass the sheep from one side to the other as they are drenched and dressed, or for removal of litter, disinfection, &c.

The bowels being usually constipated at the outset, give Aperient No. 1, page 86, to which is added 2 drs. each of powdered ginger, gentian, and carbonate of potash. Give a Clyster No. 1, page 90, in addition, so as to gain time. The mouth should be washed at least twice a day with Antiseptic Lotion No. 2, page 85, or Astringents No. 3, page 87, diluted by an equal quantity of water. As a drench, mix Diuretics No. 1, page 92, with Tonics No. 1, page 99, and give these morning and night. In examination of the feet, open the vesicles and cut off only *loose* pieces of horn from the feet, with those which may tend to prevent the escape of pus, &c. A large quantity of Antiseptic Mixture No. 8, page 85, should be at hand so as to dress every affected part by means of tow on the end of a stick, or held between the fingers; afterwards saturate tow or a rag in the mixture and secure it between the hoofs and round the fetlocks. This should be done daily. When abscesses form, open them as soon as possible and use Healing Lotion No. 4, page 96, or the Antiseptic Mixtures No. 7 or 8, page 85. If weakness is present, give nitrous ether as a stimulant, with Vegetable Tonics No. 3, page 99. Use linseed mucilage during constipation as a vehicle for the drenches; and if diarrhœa or looseness appears, give flour or starch gruel with Astringents Nos. 7 and 8, page 87, or Anodyne No. 2, page 85.

Draw the udder frequently, and smear it with the Antiseptic Mixture No. 8, page 85, and treat abscesses in the way already described. See "Mammitis," chapter xviii.

Small-pox.

The present generation knew nothing of this affection until 1847, when it was imported from Holland. The disease has continued to harass continental countries for centuries, in consequence of the movement of stock towards ports of embarkation and the frontiers of other nations for the purposes of trade. The illicit travelling of stock is also a fertile source, while a disreputable looseness may be noticed in the legislation of some Governments.

Small-pox is known by many names, but in this country we have been content to speak of it simply as we now do. We understand it to be a highly contagious fever, characterized by eruptions on the skin, readily propagated among sheep, but very seldom communicated to other animals. The poison is easily transmitted by means of the atmosphere over a distance of 500 yards, by the clothes of persons in contact with diseased animals, by the manure of the fold, and by dogs. The losses have been as high as 50 per cent. in England, but on the Continent it is characterized by a mildness which is never known when it is communicated to the stock of a clean country. The false practice of inoculation serves to keep the disease alive in many districts abroad, and therefore the greatest vigilance should be exercised in the system of importation to this country. Once introduced, in consequence of the habit of sheep to keep close together, as well as from being confined to small pastures, meeting with other flocks on the runs, or going over the same ground, the disease spreads with alarming rapidity, few escaping and many dying. Small-pox is characterized by a period of incubation which varies from seven to fourteen days, but after inoculation the signs appear as early as three days. The disease when naturally conveyed is influenced very much by temperature and association. Cold weather and isolation of the sheep greatly retard the development of the signs, whereas warm or hot weather and crowding especially hasten their appearance.

An increase of animal temperature marks the incubative stage, and this is followed by dulness and fever. The skin exhibits a "flea-bitten appearance," each spot becoming more inflamed, enlarged, and elevated, and by the eighth or tenth day is filled with a clear fluid. Somewhat later the contents become turbid, denoting the formation of pus. The colour is now yellow, the covering and contents opaque, and the surrounding skin very pale. Gradually the pustule flattens, transudation occurs, and the swelling appears more diffuse; the contents are drying up, and a brown scab forms on the top, which falls off in a few days, leaving a pit or depression in the skin.

Constitutional disturbance is very great when the eruption is extensive and *confluent*, that is running together. A purulent discharge flows from the eyes and nostrils, the breathing becomes thick, and an intense thirst tortures the animal; an exhalation almost unbearable arises from the skin, the breath is equally fœtid, the mucus membranes acquire a blue appearance, the tongue protrudes from the mouth, the animal pants, drawing breath with extreme difficulty, and death takes place about the eighth day, the eruptions rarely having passed on to the formation of lymph. In mild cases the fever is slight, and disappears as soon as the eruptions are developed, convalescence being established by the fifteenth day or thereabouts.

Irregular forms are observed, such as the non-appearance of external eruption, which denotes internal swellings, giving rise to intense fever, and death after an exceedingly offensive diarrhœa.

The mucous membranes may become the seat of the eruption, when the difficulty in breathing is great; a purulent discharge glues up the nostrils, and the animal breathes through the mouth; the tongue protrudes, and has a blue or purple colour. Disturbance of the digestive organs is denoted by fœted diarrhœa.

Sometimes the joints are affected, the hoofs slough off, large and indolent wounds succeed the eruption, the scales of which have been rubbed off during violent irritation of the skin.

At other times the vesicles are filled with blood; or gaseous accumulations appear beneath the skin, the result of a form of decomposition which characterizes most malignant blood poisons at the close of life.

The *treatment* of small-pox, in whatever part of the world it may have been undertaken, has hitherto proved alike dangerous and abortive. Like cattle plague, contagious pleuro-pneumonia, and epizoötic aphtha, it cannot be cured. It will run through its course in spite of all measures that may be devised. We know of no specific at the present day sufficiently capable of destroying the poison; therefore the exercise of all means we possess should be directed towards mitigating the sufferings of the animal, ministering to its comfort, and using every endeavour to prevent the spread to others by instant segregation on the first appearance of disease. As long as affected animals live they are powerfully manufacturing and spreading the poison, and animals or men in contact with them may assist in a wider distribution. It is, therefore, of the greatest importance to adopt a definite and immediate line of action. We believe it to be sound policy to slaughter at once all that are affected, and thus to effectually nip the disease in the bud as it were, at the same time separating the flock into small groups here and there. The other plan is to isolate all mild cases at once; destroy unreservedly all which indicate an irregular form, and bury them deeply. Inoculation serves to perpetuate the disease by keeping alive and spreading the poisons, and experience proves the fallacy of its adoption under any circumstances.

When it is determined to practise treatment of the

milder forms, let there be no communication between other animals and those affected. Appoint special men to wait upon them, and let these remain continually near them, all their wants being supplied on the spot. If these principles are carried out, small-pox may be limited to the fold or yard where it was first observed. See Disinfection at the close of this chapter.

The needful remedies are, Febrifuges No. 1 or 2, page 94, and Diuretics No. 1 or 2, page 92. Aperients No. 1, page 86,—during constipation always adding or giving alone nitrous ether as a stimulant to combat prostration. As a dressing to the discharged pustules use Antiseptics No. 6, 7, or 8, page 85, with unremitting attention. Sponge the nostrils freely with No. 6 or 7, page 85, and apply it to the whole of the fleece by means of a soft brush carried only in one direction, that in which the wool lies. Disinfect the litter freely and regularly, and when removed carry it to a suitable place for being further acted upon or summarily burned. Fumigate the air of the buildings also from time to time, or use the Antiseptic Lotions No. 2 or 6, page 85, by means of the spray producer. The fumes of burning sulphur only are to be relied upon.

Measles, or Rubeola.

This disease is of somewhat doubtful occurrence in sheep. It has been described by continental veterinarians as taking place among sheep and pigs, but others are inclined to look upon it as a modified form of small-pox. If it should visit these shores, which is just possible, as it is pronounced contagious, and produced by inoculation with the nasal discharges, we shall have an opportunity of forming judgment upon it.

Measles, or rubeola, is denoted by the appearance of catarrhal symptoms, with high symptomatic fever, swelling of the head and throat, constipation, and loss of appetite. An irregular erruption takes place about the second or third day, which is confined to the face, chest, sides of the body, abdomen, and insides of the thighs. It was observed to consist of a diffused elevation and redness, in the centre of which a hard substance was evident. Pressure also dispersed the colour, which immediately returned. At the expiration of another day a papulous eruption took place over the place of the central point of hardness, and in two or three days more the acute signs were allayed, the spots became brown, the cuticle peeled off, and the whole subsided about the eleventh day or sooner. If colic and diarrhœa set in the case usually proved fatal. A peculiar and offensive odour was exhaled from the skin.

Treatment.—Complete segregation of all affected animals. The bowels must be regulated by mild aperients, and febrifuges must be given to quiet the system, and the action of the skin promoted as much as possible.

Means of Preventing Contagion.

This is such a wide subject that we can do no more than give very brief recommendations. The subject is more fully dealt with in "The Cattle Doctor."*

All persons of intelligence acknowledge the benefits of *cleanliness* in their habitations, and the admission is extended to our domestic animals. In disease it is called for urgently and persistently as an efficient agent in alleviating, mitigating, and assisting in the cure of disease. Cleanliness greatly arrests the spread of contagion and limits the tendency to other diseases, particularly those of a widespread character. Next in order come *ventilation* and *disinfection.*

Ventilation.—Close and ill-ventilated buildings create a great tendency to disease by depriving the system of pure air, which is essential to health. Good ventilation—providing an abundance of fresh air—dilutes poisonous or contagious matters, and thus lessens their effect upon

* Illustrated by many engravings, plates, &c. Published by F. Warne & Co., 15, Bedford Street, Strand.

the animal body. This explains how contagious diseases lose their power at a distance in the open air. Strong acids possess an intense action in their concentrated form, but mixed with a great quantity of water they are not detected. Poisonous miasmata are paralysed if not destroyed in a similar manner.

Disinfection.—This is the act of destroying or neutralising the products of contagious disease by processes or substances of a chemical nature. The applications are of several kinds: *solid*, *fluid*, and *gaseous*. The first are used as dry powders, or by solution in water to form the second; the third are employed as fumigations.

1. *Chloride of lime*, so called, is perhaps one of the most efficient agents for complete disinfection. It should be thrown over the floor and matters to be purified by means of a large dredger or canister, the lid of the latter being perforated with large holes, and after lying some time the whole is carefully swept together, treated with a further addition of chloride of lime, thoroughly mixed, and removed to a proper receptacle. There are, however, great objections to the use of chloride of lime. Its strong, pungent and suffocating odour is often detrimental, especially to diseased animals. In empty buildings it may answer as one of the cheaper purifying agents, but it must also be borne in mind that it renders the manure absolutely worthless as a fertilizing substance.

2. *Chlorine gas* is a valuable purifier, as it may be caused to penetrate the "nooks and crannies" of empty buildings when the atmosphere has been for some time charged with morbific matter exhaled from the lungs, skin, etc., of diseased animals. It is prepared by the following process. A sufficient quantity of the black oxide of manganese is placed at the bottom of a clean Florence or salad oil flask, and strong muriatic acid is poured over to cover it. The flask is then fixed in a retort stand and the flame of a spirit lamp applied beneath. The acid is shortly heated to ebullition and the yellowish green chlorine gas, possessing a powerful odour, is rapidly evolved and diffused. To give full effect

to the process the doors and windows should be previously closed, all implements, harness, etc., removed, and the operator, satisfied as to the efficiency of the process, should withdraw.

Another, and certainly more simple, yet effective plan consists of fumigating by means of sulphurous acid. Four bricks are placed to form a square, into which red hot cinders, etc., are placed. Doors and windows being closed, four to eight ounces, or more, as required, of flour of sulphur is thrown on the fire, and a speedy exit is to be made. In each case the building should not be reopened for ten or twelve hours, and air must be freely admitted from all sides before any person enters.

The process of disinfection and cleansing is now greatly simplified, as well as rendered certainly effective, by the introduction of a new compound called "Sanitas," in various forms of preparation, *viz*, oil, powder, soap, etc. For an extended and constant use amongst stock the crude "Sanitas" oil may be used with pure water, poured from an ordinary watering can. Mixed with soap and water it answers for cleansing the skin, harness, woodwork, etc., etc., and the pure "Sanitas" oil (see p. 85) is a valuable application in contagious diseases to destroy the poison germs in the secretions, discharges, etc.

Moderately weak solutions of crude "Sanitas," say one part to eighty, or one hundred of water, and thoroughly mixed, are invaluable for cleansing the skin, fleece, etc., on recovery from infectious diseases. Such solutions should also be used by shepherds for washing the hands, and even their clothes, after giving assistance in difficult parturition, and also as a dressing to the genital organs, external as well as internal. The bedding and litter of the fold should be likewise sprinkled.

Disposal of Manure.—From ample experience gained in reference to this matter, we are encouraged to make a few special observations. We believe that few men and boys—attendants on sick animals—sufficiently understand the importance of disinfection; hence they are apt to dress the manure either very imperfectly or not at all. It is then carried away and mixed with the general stock, to come out at some other time and commit afresh irreparable mischief. We would earnestly impress upon owners to see the manure is thoroughly dressed as long as it remains in the fold or building, and when removed, along with litter, &c., the whole should be *burned.*

CHAPTER XIV.

Diseases of the Digestive Organs—Sporadic Aphtha or Thrush—Acute Indigestion, or Hoven—Chronic Hoven—Choking—Impaction of the rumen—Foreign bodies in the rumen—Diseases of the second stomach—Dropping the cud—Chronic indigestion—Colic—Diarrhœa—Dysentery—Enteritis, or Gastro-enteritis—Peritonitis—Congestion and inflammation of the liver, jaundice, &c.

THE diseases as well as disorders of the digestive organs of the sheep are numerous, and exceedingly fatal: not as independent affections, but in their tendency to develop other states as blood poisons. This has already been widely shown in previous as well as following chapters. As a class they are of great importance, and call for caution and extreme watchfulness.

Thrush of the Mouth.

Sporadic Aphtha.—Although this disease resembles, to a certain extent, the formidable scourge, epizoötic aphtha, the distinction is readily observed on close attention. It rarely attacks more than one or two out of a number, and younger animals are usually affected. Symptomatic fever is present at the outset, and shortly a crop of vesicles is observed upon the tongue, gums, lips, &c., accompanied by pain and inflammation, which prevent the animal feeding and lambs from sucking. A white incrustation usually succeeds the vesicles, which appears not unlike a fungoid growth to the uninitiated. A very bad form accompanies certain conditions of the digestive organs associated with blood diseases.

Treatment.—In the simple form, Aperients No. 1 or 3, page 86. Wash the mouth with weak Astringents, as No. 2 or 3, diluted with an equal part of water. Feed on soft food, or give nutritious gruel, &c. Attention must be given to the ewe's udder: if vesicles form, treat as for sore teats and garget (chapter xviii.).

Acute Indigestion.

Hoven, Hoove, Blown, Blasted, Dew Blown, Fog Sickness.—Acute indigestion consists of sudden distention of the first compartment of the stomach by gas eliminated from food. It is common after sheep get on new clover wet with dew; the late innutritious grass of autumn or early winter, wet with fog, also produces it readily. Besides, it is a common sign of other affections.

Treatment.—Strong doses of spirits of ammonia, ½ oz. in ½ pint of water, should be given. When extreme urgency marks the case, use the probang and withdraw the stilette; otherwise put the trocar and canula through the left flank, taking care to avoid the kidney.

Chronic Hoven.

Acute indigestion may assume chronic characters from frequent operation of the causes. The powers and secretions of the organs are diminished, and the inconvenience is common after food. The animal becomes poor, weak, and listless; the mouth is foul and mucous membranes yellow; bowels irregular; tone and condition lost.

Treatment.—Gentle Aperients No. 1, 2, or 3, page 86. Clysters No. 1 or 2, page 90. Tonics No. 1, 2, or 3, page 99, as required.

Choking.

When accumulation of food or the lodgement of a piece of root takes place in some part of the throat or gullet, signs of great distress follow, and animals are not infrequently lost from rupture of the stomach, and even the gullet. When the obstruction is in the throat, it may be reached by a long spoon, &c., and withdrawn; if somewhat lower, it may be caused to pass by gentle mechanical means, after a few spoonfuls of oil have been poured down. When still lower, first try the action of the oil, which should have a drachm of sulphuric ether added to every ounce. Four ounces of oil may be thus prepared, small portions being used probably with success. If these measures fail, the probang must be passed, observing great care to avoid wounding or tearing the gullet.

Impaction of the Rumen.

Maw-bound, Grain-sick.—When sheep gain access to unusual kinds of food, such as grain on the barn floor, they are liable to fill the rumen so as to occasion pain and danger from fermentation of the contents. The swelling, seen in the left flank, is felt to be soft, and leaves an impression of the fingers. The signs are slowly developed, but great distress at length marks the case.

Treatment. — Clysters every half-hour, No. 1 or 2, page 90. Strong Aperients, as No. 4, page 86, followed by 4 or 6 drm. doses of spirits of ammonia, hourly, in ½ pint of linseed mucilage. If extreme urgency is evident, the paunch must be evacuated by opening the flank.

Foreign Bodies in the Rumen.

Owing to the mode of management, sheep are not able to accumulate such a motley variety of substances within the paunch as are sometimes found in the stomach of cattle. The almost only substances found are concretions of wool and mucus, which probably pass on, if the animal is allowed to live, and are broken up and dissolved in the secretions of the fourth stomach.

Diseases of the Second Stomach.

The reticulum, or honeycomb bag, is rarely affected by independent disease; it suffers in common with others in the various forms of indigestion. When it is the subject of special disease the signs are by no means well defined: they are simply those of indigestion generally. It is only when the passage of foreign bodies takes place from the rumen to this organ, and, having sharp points, they penetrate the walls, during which inflammation is set up, and it may be followed by an abscess. This is the usual result of a nail, pin, or other substance, which, having entered by the point, can proceed no farther, and it remains, keeping up the irritation. In addition to the signs of chronic indigestion, there is high fever and vomiting, neither of which are allayed by medicine, but are persistent to the end, death terminating the case sooner or later.

Stomach Staggers.

Impaction of the Third Stomach, Vertigo, Fardel-bound.—This does not appear to be common among sheep, but occasionally, when food is scarce, and that which is given is dry and innutritious, a few individuals in a flock will be found suffering more or less from what is viewed as ordinary constipation. The signs are slowly developed, and consist of separation from the flock, dulness, suspended rumination, fever, oppressed breathing, staring and bloodshot eyes. The bowels pass only small black dejections, which are glazed and offensive; colicky pains arise, and the animal grinds the teeth, or grunts if moved. If the first stomach is distended, the sufferings are intensified, and, in walking, the sheep reels, or occasionally runs wildly about until, exhausted, he falls, and probably dies in convulsions, the brain having become implicated.

Treatment.—Strong Aperients, as No. 2 or 4, page 86. Clysters No. 1 or 2, page 90, every half-hour. Stimulants, as spirits of ammonia and ginger—½ oz. of the former with 2 drms. of the latter, every hour. If the bowels exhibit signs of irritation, substitute nitrous ether for the ammonia, and add 20 grs. of the extract of belladonna. If the head is affected, use ice, cold water, &c.

Dropping the Cud.

A simple form of indigestion, due to the effects of inferior food. On one occasion, we saw a whole flock suffering, the pasture being covered by the half-masticated grass, &c., when mangold leaves had been given. This was probably owing to the acid developed in them. Another flock also suffered in a similar manner, the bulk of the food being very inferior turnips. The disease consists of dropping the food from the mouth in remastication, none of which is returned to the stomach in well-marked cases.

Treatment.—Mild aperients, followed by vegetable tonics, perform a cure if the animals are allowed to have a fair proportion of proper food.

Chronic Indigestion.

Impaired appetite, or dyspepsia, is common to sheep when subjected to such management as alluded to in the preceding paragraph. The usual desire for food is absent, and he prefers to lick the stones, posts, &c., and, if the causes are not removed, sand, soil, and even manure, &c., will be devoured. We have frequently found earth and sand in enormous quantities in the fourth stomach and intestines. The animal soon loses condition, the bowels become irregular, and the first stomach is frequently hoven. Diarrhœa or dysentery usually terminates the existence of the sufferer, or, if he forbears gratifying the morbid appetite to the extent we have described, the foundation of other maladies will be laid

—as rickets, rheumatism, or tubercular consumption in its various forms.

Treatment.—Follow the advice given for "dropping the cud." If diarrhœa has set in, treat as under that head.

Colic.

The simple manifestation of pain in the bowels is rarely an independent disorder. It is usually a sign of other diseases. If it appears to be unassociated with any other malady, administer Anodyne No. 1, page 85.

Diarrhœa.

A discharge of fluid fæces, with straining, and more or less pain, is known as diarrhœa. It is commonly a sign of other diseases, and often ushers in their appearance. In its simple form, it is due to some irritant—as indigestible food; or foreign bodies—as sand, earth, worms, &c.; and it may result from good food when too laxative and in abundance.

Treatment.—Remove the cause. Aperient No. 1, page 86, in *half doses*, to which ¼ oz. of laudanum may be added. In older-standing cases, give Anodynes No. 2, page 85, or Astringents No. 7 or 8, page 87; for lambs, ½ or ¼ doses.

Acute Dysentery.

When simple diarrhœa is neglected, it may terminate in dysentery. Food of an acrid and indigestible kind will also develop the disease, which consists of a profuse liquid discharge of fæces, mixed with mucus, and even blood. Dysentery, like diarrhœa, is frequently the result of blood poisons. There is much straining and pain, with arched back and full abdomen. The appetite is capricious, and generally an aphthous eruption (see page 124) forms in the mouth. Thirst is intense, and the sufferer will plunge his head up to the eyes when he gains access to water, often drinking until he bursts.

Treatment.—See recommendations for the following disease.

Chronic Dysentery.

This disease lacks all the urgency seen in the acute form. The discharge from the bowels having continued in a sub-acute form from the first, wasting of the body has proceeded at some length, and emaciation is now largely present. The sheep is hide-bound, starved, and exhausted on the least movement, reeling as he walks. The wool is clapped, and hangs in harsh, dirty locks; the membranes are pale, the abdomen pendulous, and a swelling is present beneath the jaws. The dejections are offensive and bloody, and the membrane of the rectum is ulcerated. Such animals rarely recover if they are allowed to live to this stage.

Treatment.—Gentle Aperients No. 1, page 86, in half-doses, with 10 or 15 grs. of powdered opium, or opium and calomel may be prescribed, alternated with Anodyne No. 2, page 85, or Astringents No. 7 or 8, page 87. Occasionally metallic astringents may be called for; and if the intestinal mucous membrane refuses to absorb, medicines may be passed beneath the skin. Emollient and medicated clysters may also be useful, and antiseptics to correct the foul discharges. Later, Tonics No. 1, 2, or 3, page 99.

Enteritis.

Inflammation of the Intestines.—It is most likely that the fourth stomach participates in the inflammation which attacks the intestines. It should, therefore, be called *Gastro-enteritis.* Inflammation of the stomach and intestines arises in consequence of acrid and irritant vegetation being taken as food, and from this circumstance, as well as its usual prevalence over a district, it has been named *Enzoötic Dysentery.* The symptoms are gradually developed. When, however, metallic or other chemical poisons have been swallowed, the signs are sudden and acute. As an ordinary affection the manifestations are somewhat as follow:—The appetite is morbid and variable, and rumination is suspended; the secretions are checked, and bowels constipated. Intense symptomatic fever is now set up, colic and hoven are frequent, thirst is intolerable, urine and fæces are offensive and mixed with blood; straining is usually severe, and the sufferer stands with arched back, bleating, in pain and grinding the teeth incessantly; weakness and general prostration gradually succeed, and death at variable periods from one to two weeks. Sometimes an offensive diarrhœa sets in previously, but insensibility and convulsions may precede death in either case.

Treatment.—Aperients No. 1, page 86, to which belladonna should be added. Clysters No. 1, page 90. Allow tepid water or linseed mucilage as a drink. Febrifuges No. 1 or 2, page 94. When an offensive diarrhœa sets in, treat as stated for dysentery. Combat weakness by means of stimulants with belladonna, and provide for the comfort of the animal as much as possible.

Peritonitis.

Inflammation of the lining membrane of the abdomen, &c.—This is not common. The usual causes are direct injuries, or operations as castration. The disease is characterized by much pain, symptomatic fever, and unequal temperature. The sufferer frequently puts his nose to the flank, crouches and lies down, but soon rises; he also bleats and grinds the teeth. Later the acute symptoms somewhat suddenly subside, when the abdomen swells, weakness increases, and at length he lies down, dying in four or five days after being seized.

Treatment.—Aperients No. 1, page 86, to which aloes may be added. Clysters No. 1, page 90, until the bowels have been satisfactorily moved. Febrifuges No. 1 or 2, page 94, or calomel and opium, &c. Blisters to the abdomen, page 87.

Jaundice, or the Yellows.

So called in consequence of the orange-yellow colour of the membranes and skin, &c. This condition is not to be regarded as a disease in itself; it is merely a symptom of several disorders of the digestive system. In its simplest form it arises from any obstruction which may impede the flow of the bile towards the intestines, a condition which is observed in frequent instances. There are always attendant signs of dyspepsia—slow pulse, and languor or tendency to drowsiness or sleep. Jaundice may arise from continued or frequent congestion of the liver, when it will assume a more or less persistent and chronic form. The causes are irregular feeding and want of proper exercise; and derangement of this organ may so interfere with the normal functions of nutrition, as well as alter the constitution of the blood, as to render the system prone to other affections, blood diseases especially. We have shown how far this is the case when alluding to the nature of enzoötic diseases.

Treatment.—Aperients, with calomel, No. 3 or 4, page 86, to produce a regular action of the bowels. Clysters No. 1 or 2, page 90. In long-standing cases, calomel and opium; iodide of potassium, neutral salts under veterinary supervision. When convalescence is established, Vegetable Tonics No. 2, page 99. Regular exercise is indispensable.

Inflammation of the Liver.

Acute febrile signs, and pain or pressure on the right side, are important characteristics. Jaundice is not necessarily present. Unusual coldness of the horns, ears, and extremities, elevation of temperature only slight, pulse slow and infrequent, respiration slow and abdominal. The bowels, at first loose, become constipated, and the fæces are glazy, and impart a yellowish-green stain to paper; the urine is very deeply coloured, and the wool is dry and harsh. Colicky pains come on, the animal staggers, and fainting fits occur. The peritoneum is apt to partake of the inflammation, when dropsy follows, and death after great emaciation.

Treatment.—Aperients and clysters, as advised for jaundice. Blisters to the right side.

CHAPTER XV.

Enzoötic or Blood Diseases caused by Animal Poisons, non-contagious, but producing a putrid fever in other animals by direct inoculation—Carbuncular erysipelas or black-quarter—Splenic apoplexy—Gloss-Anthrax—Braxy—Pre-parturient apoplexy—Heaving or after-pains, or parturition fever in ewes.

THE diseases included under the above head form another section of the large class of indigenous or enzoötic maladies, which, owing to a state of soil, mode of cultivation, and climate, have prevailed for generations, and having completely put a stop to breeding in many districts, have likewise reduced many farmers to bankruptcy and ruin.

They are due to over-fertility of soils, the sudden influx of rich grasses caused by warm rains and genial weather, after confinement to bare and innutritious pastures, and to a system of forcing by rich nitrogenous food during prolonged idleness and confinement after a hard winter. These are causes which give rise to an unnatural plethora, a state in which the blood material is abundant, but not properly elaborated. An excess of such material gives rise to sanguineous extravasation, besides undergoing a process of degeneracy within the blood vessels.

Another fertile source of disease is the excess of manurial elements in the soil. This produces vegetation more or less impregnated with septic materials or organisms, which find their way to the blood, and not only disturb its functions, but destroy its constitution. The bodies of animals dying from these affections are frequently found to be already undergoing the process of putrefaction. Home pastures and rich marshy ground are dangerous places for feeding, unless great caution is observed.

Carbuncular Erysipelas.

Black-Quarter or Black-Spauld, otherwise known as *Quarter-Evil*, *Quarter Ill*, *Joint Ill*, *Black-leg*, *Speed*, *Hasty*, *Puck*, *Sheet or Shoot of Blood*, the *Blood* and *Blood Striking*, and *Inflammatory Fever* by Youatt.—This is an exceedingly rapid and fatal disease. The first intimation of its presence in a flock is the probable discovery of one or more dead carcases at the first morning visit. Animals in a rapidly thriving condition, especially after faring badly for some time, suddenly gaining access to rich food or pastures, are the first to suffer. The disease is still more rapid if, added to this, the sheep are confined to small folds or pastures, and thus take too little exercise for the necessary stimulus to healthy function. The practice of feeding animals in a kind of spasmodic manner—starving at one period and cramming at another—acting under a prevailing impression that young animals do not need good food, that it is time enough to give cake, &c., when they are wanted for the butcher, happily prevails less than in past years. But the sheep farmer has yet to contend with the vicissitudes of season, soil, and climate, and these produce such changes in his crops as well as constitution of his animals that he cannot securely withstand them. He can, however, pay more attention to his lambs, and by observing the rule to insure a sharp appetite, good food, and certain exercise, he will save hundreds of lives as well as pounds.

The disease is preceded by an unusual activity of all the functions, a peculiar liveliness, and staring appearance—signs rarely noticed, as they may happen towards nightfall. In a very few hours affected animals are seen away from the flock with hanging ears, eyes half closed, head held down, and declining to move. Probably one leg is already flexed and the foot held up or resting on the toe, and he is lame when disturbed. Symptomatic fever now sets in severely; large diffused swellings occur on the back, hips, quarters or limbs, which on pressure emit a crackling sound. Prostration ensues, the animal lies or falls down, the paunch swells, eyes and tongue protrude, foam issues from the mouth and nostrils, insensibility and convulsions follow, and death in a few hours. Milder cases may extend to a day or two, and, under certain conditions, large wounds form over the body, the animal lingers for several weeks, and eventually may recover. This, however, is very rare.

Treatment is seldom required owing to the rapidity with which the disease runs its course. The secret of safety lies in *prevention.* If the state of the animal encourages the use of medicines, give Aperients No. 2, page 86, with which nitrous ether is blended, and repeat the stimulant alone every three hours in linseed mucilage. Clysters Nos. 1 and 2, page 90. Astringents to the mouth No. 2 or 3, with a pint of water added. Open abscesses freely and early, and use Antiseptics No. 2, 3, 6, 7, or 8, page 85. After the subsidence of irritation give Tonics No. 1, 2, or 3, page 99.

Prevention usually consists of bleeding* and purging the apparently healthy; a change of pasture and food must be given, and a certain amount of enforced exercise insured, either by putting them on a bare pasture, or driving them to or from a distant one night and morning. Instead of blood-letting and frequent purging, which is troublesome and not always successful, besides being somewhat contrary to requirements, we have preference for a course of neutral salts and tonics after a laxative, under the supervision of a veterinarian.

N.B.—The carcases of all animals slaughtered, and those dying in consequence of this disease, should be disposed of only by burial or burning.

* In all cases where bleeding is called for the jugular vein should be opened. Cutting the tails and ears is useless and barbarous.

Splenic Apoplexy.

This disease is usually confused with carbuncular erysipelas, and described under the same term as "Blood striking" in nearly all the existing works on diseases of the sheep. To avoid this confusion we prefer to speak of them under the respective medical terms which denote their true nature, avoiding every other, except to enable the reader to identify the malady. Splenic apoplexy is a blood disease, common to mature animals as a rule. It is due to a morbid animal poison developed by an unnatural plethora, or imbibed in the food when septic organisms are liberated from rich pastures, &c. As the term implies, the local morbid signs are discovered in enlargement, turgescence, and often rupture of the spleen. Splenic apoplexy runs its course even more rapidly than carbuncular erysipelas. The early signs are also almost identical, as sudden deaths, and in the living unusual excitement, &c., combined with constitutional disturbance, constipation, and staggering gait. These signs are soon aggravated. The breathing is loud and stertorous, fever runs very high, the pulse is imperceptible, and the sheep falls, seized by convulsions; bloody froth escapes from the nostrils, and the bleating is mournful; he dies at variable periods within a few hours from the attack.

Treatment.—The same remarks apply here as to carbuncular erysipelas. It is rare that recovery takes place; nevertheless when an opportunity is given and the animal is a valuable one, it may be advisable to attempt to restore him. An oleaginous aperient, No. 2, page 86, with a dose of nitrous ether, should be given at once, the latter being continued every three-quarters of an hour until strength returns. The propriety of bleeding is doubtful; much depends upon the state of the pulse, &c., when the animal is found.

N.B.—Blood and mucus of diseased animals will convey the often-fatal malady to man, known as malignant pustule.

Preventive measures are the same as advised for carbuncular erysipelas.

Gloss Anthrax, or Blain.

This form of anthrax fever is characterized by rapid swelling of the tongue and fauces, &c., with development of pustules and malignant carbuncle. The disease does not appear to be contagious as far as the atmosphere is concerned, but, like splenic apoplexy, the blood, mucus, &c., will convey a putrid fever of the nature of malignant pustule. Gloss anthrax, as a rule, prevails when other diseases, as epizoötic aphtha, are rife. In addition to the foregoing signs symptomatic fever is severe, to which succeed prostration, insensibility, and death. The nose, face, and neck swell rapidly, producing difficulty in breathing, and ulceration with tumefaction of the mouth cause much suffering.

Treatment.—Early bleeding, if possible, from the jugular vein, oleaginous purgatives and stimulants if the animal is able to swallow; antiseptic and astringent washes to the mouth, followed by stimulants, mineral acids, tincture of steel, &c.

Braxy.

Two forms of this affection are recognised by the shepherd, constipation and diarrhœa being the distinguishing marks. With regard to them, we may observe that only one form of disease really exists, diarrhœa being dependent upon certain modifications which may ensue. Braxy is a blood disease similar in nature to carbuncular erysipelas, and is often more fatal. Sometimes the fleece may be observed beforehand to be very dry, and is "fallen" or "clapped," and in walking the animal takes a short step. Bloodshot eyes, an excited appearance, high febrile symptoms, constipation, and highly coloured urine. The animal separates from others, and droops the head and ears. If diarrhœa is present the paunch swells, but in constipation there may be straining, and the stools are hard, black, and highly glazed, or covered with mucus and blood. More frequently the animals are found dead, having fallen into water or over precipices, leading to the assumption that both have been accidental and independent accidents.

Treatment.—This should be essentially preventive, based on simple dietetic principles. Aperients No. 2 or 4, page 86, followed by continued stimulants. Clysters during constipation, No. 1 or 2, page 90. The whole flock must be seen to. They must be confined less to the fold, the diet reduced, and more exercise given.

Sheep in the last stage of Braxy.

Anthracoid Diseases.

These are diseases which are supposed to partake of the characters of anthrax in a modified degree. Notwithstanding we have yet to learn more in this particular before we are able to act safely with regard to their cure and prevention.

Pre-Parturient Apoplexy.

Apoplexy before Parturition.—The errors of diet in breeding ewes are not more apparent in any other disease of the sheep. The affected animals stray away from the flock, are dull, lie down, toss or jerk the head upwards, and grind the teeth. When suddenly disturbed they jump wildly, run forward holding the head upwards, lifting the feet high, and eventually fall head foremost. They appear stunned, but after a time rise, walk stiffly to a distance, lie down, and become dull and sleepy. The appetite is lost, rumination has ceased, the belly is tucked up, the ability to rise and walk becomes less, the eyes become glassy, blindness and insensibility, sometimes with convulsions, follow, and the animal soon dies. These signs appear at variable periods prior to lambing, and continue as long as five days. Those affected, as a rule, have twin lambs or triplets in the womb.

Treatment.—This disease essentially calls for prevention by care in feeding pregnant ewes. The timely remedies are probably a moderate blood-letting, followed by aperients and salines. Common or rock salt as a condiment must always be avoided, except where medical opinion exists to the contrary.

Heaving, or After Pains.

Inflammation, or the true Parturition Fever of Ewes.—This affection usually manifests itself about the second or third day after lambing, and is known further by the animals straying away from the rest of the flock, having no inclination for food, and remaining in one spot. The flanks are full, head and ears are drooping, symptomatic fever is present, and frothy mucus adheres to the angles of the mouth. The animal pants and strains, the external genital organs are usually swollen, hot, and red, and are probably discharging a dark and offensive fluid, but afterwards become deeper coloured, finally purple and black. Straining continues and exhaustion increases; she moves with difficulty, and has forsaken her lamb. The urine is expelled with difficulty, and is ammoniacal. Dulness is succeeded by insensibility, mortification having set in, and death speedily follows. Those ewes affected have mostly more than one lamb, and were esteemed among the most thriving.

Treatment.—Put every ewe by herself, if possible, some distance from the rest, and the shepherd in attendance should put on a special frock while caring for the diseased. He must on no account handle an apparently healthy ewe without having first washed his hands and changed his frock and boots. It is by neglect of these precautions that the disease proves so widespread at times.

The affected animal should receive early attention. An oleaginous aperient with a stimulant should be given at once, and the stimulant repeated hourly until the patient rallies. The external and internal organs of generation should be carefully cleansed by the Antiseptic mixtures, as Nos. 2, 6, 7, and 8. It is good practice to inject half a wineglassful of the latter direct into the uterus at the close of the operation morning and night. At a later period, neutral salts are required.

The shepherd should be supplied with an ample supply of the Lotions No. 2, 6, or 7, in which his hands must be washed after attending to each ewe; besides which, his frock, &c., should be hung in an outhouse during the day, and subjected to fumes of burning sulphur, or frequently washed in strong alkaline solutions, and afterwards wrung out of Lotion No. 7. The fleece of each affected ewe also must be similarly treated with the lotions named, particularly before going with the rest of the flock.

The antiseptic treatment of this malady has taken great hold upon certain minds during the past few years as a new discovery. It has been long known that antiseptics have had a beneficial action in this disease, and, moreover, veterinary surgeons have recognised the great tendency for the disease to spread when due care is not observed. We remember during our pupilage, as far back as 1845, making up gallons of a mixture known as "lambing oils," their action being that of the antiseptic character. Clater, in his edition of his "Cattle Doctor," published in 1810, gives a prescription essentially antiseptic, and particularly recommends the mixture being carried to the "matrix or womb, either by the hand or with a spoon." It is thus with half the information which veterinary science develops: there are many on the look out for personal gain and glory, and they strut about in the plumes of others, totally unaware how unsuited and badly fitting they are. We remember once giving a man a simple formula for a mixture, telling him at the same time it was invaluable as a healing fluid for wounds. In a few weeks we found him travelling the fairs and markets with the same as a new and important discovery. For the sake of many animals who suffer by cruel experimentation with many nostrums now sold, as well as for the saving of money now uselessly squandered, we wish there was a little more solid learning abroad.

CHAPTER XVI.

Blood Diseases which prevail as Enzoötics—Malignant catarrh—Arthritis, or joint diseases of young lambs—Red or black water—Malignant sore throat—Enzoötic typhoid catarrh, or influenza—Sanguineous dropsy or red water—Navel ill.

THE diseases we are now about to consider form an important section of the large class of blood diseases. As far as they have been observed among sheep, they prevail, as among cattle, in the form of enzoötics, *i.e.*, numbers of animals in the flocks of whole districts suffer, the causes being widespread, but nevertheless confined to the districts. The usual features are the attacks being simultaneous, or so near together as to denote the operation of the same causes upon the whole. Their disappearance is also as rapid as the origin is sudden, a change of temperature, food, or pasture, &c., being only necessary to arrest the attack. The disease spreads no farther than the limits or area in which it originated, notwithstanding the amount of inter-communication which may be carried on with flocks not already affected. The maladies are not infectious: this is evident from the numbers in the same flock which escape; but under aggravated states are capable of conveying a putrid blood-poison to healthy animals or the human subject by inoculation. Some care is therefore required in dealing with them. Possibly, on further investigation, we may be able to establish a connection with other maladies, or at least a dependence upon them, by the discovery in certain cases of particular poison organisms in the blood.

A few of the maladies which we propose to notice briefly in this and other parts of the work have not been hitherto definitely observed in sheep, or they have been described under other names. This deficiency shows how much need there is for the encouragement of sound scientific and veterinary supervision of the flocks of the

United Kingdom. If they could be submitted to such from this moment by resident veterinarians, their whole time being given up to the farms and flocks of a prescribed district, the science of sheep pathology might be considerably enriched. It would be a proud day for the stockowners of Great Britain, the eyes of our legislature being opened to the restriction of importation to dead carcases, if the army of veterinary inspectors could be drafted to our rural districts, and the staff of the whole department, as it now exists, devoted to the work of investigation and prevention of sheep diseases. How much more cheerfully we would submit to the present taxation for the support of the staff if we were assured, as we should be, that the stock owner and breeder were deriving the full benefit by a preservation of their animals, and we healthy food at a much less cost?*

Malignant Catarrh.

This disease has only been considered common to the ox tribe, and accordingly received the erroneous title of "Glanders of the ox." From reports which have reached us from time to time, from home as well as foreign countries, the symptoms of maladies given bear a close resemblance to this affection. Without claiming any identity, we nevertheless venture to allude to it. The brief description we give will at least assist in calling attention to a necessity for further observation and publication of the records.

Malignant catarrh consists of inflammation of the nasal passages and sinuses of the head, attended by copious offensive discharges of mucus and blood, swelling of the bones of the head and face, fœtid breath, troublesome cough, symptomatic fever, prostration, diarrhœa, and death. So far these signs have been seen in sheep during cold, wet weather in spring or autumn.

* This subject has been somewhat fully discussed in chapter xxvii. of "Cattle, their varieties in Health and Disease," uniform with the present work. London: F. Warne & Co.

In cattle the disease goes on to sloughing of the mucous membrane, and even the horns and hoofs. So malignant is its nature that it has been confounded with purpura hæmorrhagica and rinderpest.

Treatment.—Cooling and evaporating lotions, Nos. 1 and 2, page 96, or cold water to the head. Clysters No. 1, page 90, frequently used; and if the bowels are acutely constipated, Aperients No. 1, 2, or 3, page 86. Antiseptic fluids, No. 2, 6, or 7, should be passed into the sinuses and air passages, as well as used to ulcers and sloughing. Afterwards treat by means of ethers, turpentine, mineral acids, tincture of steel, quassia, &c.

Arthritis,

Or Joint Disease in Lambs.—This troublesome disease appears in two forms—rheumatic and scrofulous—generally when the lambs are a few weeks old; sometimes they are affected at birth. Notwithstanding that it may, in both instances, be traceable to hereditary taint, the active cause is stinting the ewes before and after conception. As a rheumatic affection, it flies from joint to joint, giving rise to great pain, lameness, and constitutional disturbance, and after some time, if the lamb lives out the disease, terminates in stiffness of the joint. In the scrofulous form the disease has a preference for the porous ends of the bone. The structures there are favourable for deposition of scrofulous material, inflammation, swelling, and subsequent abscess, which successively follow each other in the order given.

Treatment.—Perfect stillness is essential, the joint being fixed by starch bandages. Internally, Febrifuges No. 2, page 94, in one-eighth, one-sixth, or quarter doses. Embrocation to the joints, No. 1, page 92, during the acute stage, or Blisters No. 1, 2, or 3, page 87. Afterwards, Embrocations No. 4 or 5, and internally, iodide of iron.

Red or Black Water.

Asthenic Hæmaturia.—This disease exists among a number of sheep as a result of scarcity of good food generally, that to which they have access being coarse, rank, acrimonious, and indigestible, the growth of poor and undrained land. Owing to the same causes, added to which is the drain upon the system during pregnancy, the disease will sometimes appear shortly after parturition. It is usually associated with poverty and bad management. Long-standing weakness produces an imperfect assimilation and digestion, and if, added to this, innutritious and indigestible food only is supplied, very improper materials find their way to the blood, "adding fuel to the fire" and further derangement to the digestive organs. Thus we have a train of symptoms somewhat of the following kind:—Weakness, emaciation, and staggering gait; anæmic palpitation; diarrhœa, alternating with constipation; disturbed respiration, general coldness, pale membranes, sometimes intense thirst, colic, prostration, sinking, and death. The colour of the urine varies from a red solution to a black and opaque muddy fluid. The odour is also strong in proportion to the colour, that of the black urine being simply intolerable.

Treatment.—Aperients No. 1, 2, or 3, page 86; Clysters No. 1 or 2, during constipation. Internally, Astringents No. 7 or 8, page 87. Afterwards the mineral acids, tincture of steel, quassia, &c., alternated with Vegetable Tonics No. 3, page 99. Linseed mucilage forms the most suitable drink. An entire change of food must be allowed, with every means to insure the proper kind as to nutrition and digestibility.

Malignant Sore Throat.

A form of disease usually called "influenza," but erroneously. It is exceedingly fatal unless considerable skill and promptness are exercised at the commencement. The throat is swollen externally, and signs of catarrh are present. The mucous membranes assume a purple colour, the fauces are swollen, and as the external enlargement proceeds, the air passages are closed, whereby the breathing becomes difficult, and swallowing impossible. In subsequent stages the breath is fœtid, the tongue protrudes, suffocation is imminent and aggravated by a convulsive cough. The nostrils are more or less closed by a muco-purulent discharge, which adds to the embarrassment, interrupting circulation and producing dulness, listlessness, insensibility, and then death.

Treatment.—If the animal can swallow, give an Aperient, No. 1 or 2, page 86, and follow with Clysters No. 1 or 2, page 90, three or four times during the day; afterwards mineral acids, &c. If swallowing is impossible and breathing difficult, the windpipe should be opened and a proper tube inserted. As medicines use Electuaries No. 1 or 2, page 92, or administer remedies by the hypodermic method, *i.e.* beneath the skin. As dressings for the mouth use Antiseptics No. 1, 2, 3, or 6, page 85, or Astringents No. 2 or 3, doubling the quantity of water there recommended.

N.B.—The flesh of sheep slaughtered when suffering from the diseases described in this chapter, is not suitable for human food. In certain instances the blood and secretions partake of poisonous properties, and butchers flaying the bodies have suffered acutely, others even dying from imbibing the poison. The carcases should be buried deeply, and, if possible, cut to pieces to facilitate decomposition. They would be more safely disposed of by burning.

Enzootic Typhoid Catarrh.

Commonly termed *Influenza.* A peculiar form of catarrh has been noticed among sheep, and in consequence of its great resemblance to the influenza common to horses, it has been termed such in accordance. The affection is characterized from the first by a great prostration of strength, and remarkable tendency to internal congestions, particularly of the lungs, which speedily carry off the patient. In those forms marked by less acute character, debility is very prominent, the animal being disposed to stand with back arched and propping or resting against some near object. The mucous membranes exhibit the dark or purple colour, the eyes are closed, and the sinuses of the head as well as nostrils being implicated, an offensive discharge flows of a muco-purulent kind. There are indications of brain disturbance such as partial coma and dulness, and the tendency to move in a circle, arising from defective blood. Besides these, the signs of pneumonia and pleurisy, with disease of the liver, are more or less present. From the extreme prevalence in some seasons, the disease has been considered to be infectious. Whether this be a correct inference or not, is undecided. There is no doubt whatever that the functions of assimilation and elaboration of blood material are seriously interfered with, probably owing to exposure and other causes of a similar character. As it prevails chiefly in spring, when the variations of temperature are great, and particularly on wet marshy soils, we may also look for causes in the vegetation, which conveys to the system elements capable of disturbing the course of true blood formation. This view is strengthened by the fact that diarrhœa of a somewhat offensive character rather speedily sets in.

Treatment.—Move the bowels, if constipated, by a gentle aperient, No. 1, page 86. When diarrhœa is present reduce the dose one-half, adding tincture of opium one or two drams, and nitrous ether half an ounce. This treatment should be followed by stimulants and vegetable tonics, directed by the advice of a veterinary surgeon. The sheep should be placed at once under shelter, such as a shed open to the lee side, or sufficient accommodation may be improvised by means of hurdles, poles, and rick covers, &c. The apparently healthy should be removed also to a protected situation, and receive a laxative combined with a stimulant, followed by vegetable tonics, and supplied with oats, &c., as trough food. As they are attacked they should be removed to the temporary hospital referred to above.

Sanguineous Abdominal Dropsy.

Commonly known as *Red Water.* This is another form of supposed anthracoid disease, due to the effects of irregular feeding and management. It consists of a derangement of the blood, which secures a sudden effusion of a red-coloured serum, or fluid, within the abdomen. Ewes and lambs are subject to it, and the latter are sometimes dropped before time, being "water-bellied," in the language of the shepherd. Few of the

leading signs are observed, as the animals are usually found dead. They consist of absence of appetite and rumination; dulness, with disinclination to move; and a reeling or staggering gait, with stiffness. The bowels are constipated, and other secretions are checked. Finally, the eyes are staring, pupils dilated, the head is carried to one side, and the animal is probably blind; weakness and paralysis follow, with speedy death.

Treatment is seldom called for. The attention should be devoted to saving others after the first death has announced the operation of the causes.

Blood Disease in Lambs.

Usually known as *Navel Ill.* We recognise in this an affection also due to improper feeding before and during the time of conception, as well as after parturition. In some seasons, when food proves abundant, the lambs die off by scores. Care should be exercised in putting on the skins of dead lambs at this time, as the deaths are multiplied, and the cause supposed to be the original disease.

The finest lambs are usually affected. The disease is shown by dulness, sudden staggering, with prostration, bloodshot eyes, and congested membranes. The animal pants, is unable to stand, but will attempt to suck, probably, if held up for the purpose. The secretions are checked; swellings appear on various parts of the body, particularly on the neck, or throat, and at the navel, which are soft and fluctuating. Exhaustion proceeds, and death may occur in a few hours, or be delayed, perhaps, to a few days, or even a week.

Treatment.—This, and the affection named in the foregoing paragraph, can only be met by prompt preventive measures, which should be applied to the whole of the apparently healthy lambs. They properly belong to the period of gestation, when the ewes should have more regular treatment, more exercise, and less of the rich food, whatever it may be. The affected animals should be treated likewise; a bare pasture, with a course of neutral salts under the advice of a veterinary surgeon, is the only suitable plan.

CHAPTER XVII.

Diseases of the eye—Simple ophthalmia—Iritis and retinitis—Staphyloma—Fungus hæmatodes—Removing the hacks.

INFLAMMATION of the structures of the eye is common to the sheep. Occasionally it is limited to one or a few, being dependent upon local causes or accidents, &c., and in other instances it assumes the widespread nature of an enzoötic. As far as our experience goes, it is not known to take on what are called "specific" characters, as is observed in horses and cattle. This, we can imagine, may be in consequence of the protection which sheep so much enjoy in the fold. The comparatively short period of their existence, also, cuts off many of the maladies to which they would become liable, particularly under management which even now renders their short lives prematurely shorter.

Simple Ophthalmia.

In this form the inflammatory process is confined to the superficial layers in front of the eyeball. These form the cornea, or transparent portion, through which vision is effected. It is also known as *conjunctibitis*, *corneitis*, and *superficial ophthalmia*. The animal cannot bear the light, and tears flow copiously from the closed eyelids. The attendant pain is great, and an examination is made with difficulty, as the sheep not only shrinks from it, but he retracts the eyeball within the orbit, and thus it is immediately covered by the haw. The veterinary surgeon, however, by means of tact, discovers the cornea is opaque, and exhibits a bluish grey, or an opal appearance, and the inner surfaces of the eyelids are deeply inflamed. Sympathetic fever is also present. The causes are the cold winds of autumn and spring, especially if the weather is wet, the animal imperfectly sheltered, and food of a proper kind runs short. These produce an enzoötic form of the disease; but one most common, and confined to one or only a few animals, is traceable to local causes—as blows, or the insinuation of dirt, grit, and husks of grain.

Treatment.—Confine the animals to a dark apartment, but allow plenty of fresh air without draughts. If this cannot be done, blindfold them by means of opaque coverings kept constantly wet with cold water or Lotions No. 2 or 3, page 96. All foreign bodies should be removed at once by means of forceps, &c. Aperients, No. 2 or 3, page 86. Scarification of the eyelids, &c.

Iritis and Retinitis.

Inflammation of the iris or retina, and sometimes both, is often met with as a sequel of simple ophthalmia. By delay and aggravation the morbid process goes on, and eventually seizes the internal structures of the eyeball, finally producing total blindness from cataract, or destroying the functions of the retina, &c. The agony of the sufferer is intense. When the eyelids are separated the cornea is more or less transparent, but the internal humours are hazy, muddy, or semi-opaque, and sometimes a reddish shade or tinge is evident. This may also assume a yellowish appearance. When the iris is inflamed, the edges are ragged and it is stationary. The crystalline lens may be already opaque. This disease may accompany the rheumatic form of catarrh, &c., when the vicissitudes of a low temperature with wet suddenly follow warm weather. We have seen it in very dry seasons, the summer of 1868 being remarkable, as a result of intense radiation of the sun's rays from the heated ground destitute of vegetation. We hardly know of a more distressing sight than to witness a large flock of sheep, the members of which are totally blind, wandering about, running against hurdles, posts, &c., or falling headlong into ditches, rivers, &c.

Treatment.—The same as advocated for simple ophthalmia. The veterinary surgeon will probably supplement it by measures which are out of the power of the amateur to apply.

Staphyloma.

Bulging of the Cornea.—A disease which usually follows simple ophthalmia in young and sickly, or old and badly managed, sheep. The inflammation assumes a chronic or subacute character, and one or more external layers of the cornea undergo ulceration. This admits of the humours pressing forward the internal membrane, and it appears on the outside as a bluish-grey tumour, not unlike a grape. In some seasons this may prevail as an enzoötic, and occasionally the bulging leads to evacuation of the contents of the eyeball.

Treatment, as for simple ophthalmia. The veterinarian should be consulted for prevailing cases.

Fungus Hæmatodes.

Blood Fungus, or Bleeding Cancer of the Orbit.—This loathsome disease is liable to arise from repeated attacks of ophthalmia. It is a local manifestation of a state of blood disease. It is not so common in sheep as in cattle, simply because the former are not permitted to live so long. The humour, as it enlarges from within, continues to push aside the eyeball, and at length involves it in the disease, finally assuming a large irregular-shaped mass, resembling coagulated blood. Hæmorrhage is sometimes considerable; this, with sympathetic fever, causes the animal to lose condition, and if no relief is given, death follows.

Treatment.—In the early stages the knife and caustics are required; later, extirpation of the whole fungoid mass from the orbit.

Removing of the Haw, or Hacks.

This is an operation neither called for nor justifiable. The haw is provided by nature to shut out the light more effectually—to remove foreign bodies from the eyeball, and thus answers the same purpose to the lower animals as fingers do to the human subject. The haw may be swollen in inflammation of the eye,—in fact, it usually sympathises in all forms of irritation of that organ; but nothing short of fungoid growth can warrant the removal of a most useful and effective arrangement, as this proves to be. We would remind officious and ignorant operators, that on evidence of their work of this kind being produced before a magistrate they are liable to fine or imprisonment.

CHAPTER XVIII.

Diseases of the generative organs—Abortion and premature labour—Natural labour—Cleansing—Flooding—Vaginal Hæmorrhage—False or unnatural positions of the Lamb, &c.—Inversion of the vagina—Inversion of the womb—Rupture of the womb—Vaginitis—Urethritis—Peritonitis and dropsy of parturition—Garget, or inflammation of the udder—Breaking down.

Abortion.

The act of giving birth to a lamb before it is capable of carrying on the functions of life, is termed an abortion. This disease is one of the many which has ruined many breeders, and caused the reduction of sheep in the United Kingdom. It is to be prevented by a study and removal of the causes. The common names by which it is known are *slinking* or *slipping the lamb*, *warping*, *sauntering*, &c. The danger attending the accident is proportionate to the stage of gestation at which the ewe has arrived, as well as development and number of the lambs. The later the period, the greater is the risk.

The causes are as follow:—Food which is too rich and stimulating, especially when combined with want of exercise, produces an unnatural and dangerous *plethora* fatal to the lamb within the womb. *Common salt* as a condiment often proves powerful as a poison. *Bad feeding* reduces the power of the ewe to nourish the lamb. *Italian, Rye, and other seeding grasses* in pastures late in the season are apt to become ergotised; they should be mown before autumn. *Debility*, from whatever cause, acts similarly to want of food and bad feeding. *Injuries*, as being frightened or chased by dogs; these, with *running*, *rolling*, or *leaping*, produce a detachment of the membranes, thus cutting off nutrition. We recently met with a flock of sheep in which every ewe aborted. This untoward event took place in consequence of their being chased by dogs, and in sudden fright they rolled down a steep bank. *Blood Diseases* affecting the ewe prove fatal to the lamb. *Bad Smells*, besides producing excitement, and causing the ewes to run wildly about, as they arise from the putrefaction of animal remains, convey to the maternal blood microscopical organisms. These, with *ergotism* of grass, prove fertile causes which largely sap the breeder's profits. Greater care is required in the disposal of carcases of animals. They should be examined, and buried far away from other cattle; ground stained with blood and secretions should be carefully disinfected. *Boiling the carcases* beforehand, or, what is better, *burning* them with the manure, &c., would effectually do away with the serious results which so often arise from putrefying flesh on various farms.

Premature Labour

Is dependent upon similar causes as the foregoing. The danger incidental to it is due to the want of preparation on the part of the mother, as well as frequent malposition of the fœtus. Lambs born alive do not usually possess the necessary vigour. They are more commonly dead, though extracted with less delay and difficulty than the calf; and consequently there is not always the same danger to the ewe as is sustained by the cow under similar circumstances.

Natural Labour.

This depends on the ewe going the full time in gestation, and bringing forth her lamb with safety. The proper position for delivery is that in which the fore feet first appear in the birth passage, dilated and lubricated by the previous progress of the water bag, or membranes. As the feet lie in close approximation parallel with each other, the head rests in the hollow between the two shank bones. Further progress being made, the head is shortly delivered; the passage is thereby dilated, gives exit to the shoulders, and finally the whole body. There are, however variations to this procedure, but delivery is without the difficulties and dangers which attend similar conditions in the cow. In the ewe the birth-passage is comparatively much larger, and the lamb considerably more pliant than the calf. With moderate care in assisted delivery, we have been able to extract lambs in almost any position. In affording assistance to the ewe great care is required in deciding on the existence of more than one lamb; it is a sad and often fatal proceeding to attempt to extract them without this knowledge, as force may be applied to both at the same time. Thus both mother and offspring are lost.

Cleansing

Is the due expulsion of the membranes, or "afterbirth." An early and spontaneous removal within a few hours after delivery is regarded as favourable. If they are retained beyond the second day they are apt to produce inconvenience and even blood poisoning. Care is required in order to sever them from their peculiar attachments in the womb. The ewe should receive Vegetable Tonics No 3, page 99, and it may be advisable to wash out the uterus by means of the pump, using tepid water, having an admixture of the Antiseptic Lotion No. 7, page 85.

After Pains.

See "Heaving, or After Pains," chapter xv., page 139.

Flooding.

Hæmorrhage, or Bleeding from the Womb.—This sad accident may follow abortion, premature birth, and retention of the membranes. The blood accumulates more or less in the womb, and is expelled in large quantities and with painful efforts. The usual means to arrest the bleeding are cold water dashed upon the loins, or injected within the womb. A soft handkerchief may be passed into the uterus as a plug, where it should remain some time, or until further assistance arrives.

Vaginal Hæmorrhage.

Bleeding from the birth-passage often arises in conjunction with parturition by injuries inflicted by the lamb, instruments, or the finger nails. The quantity which flows is small, and is usually arrested by cold water or astringent solutions.

False or Unnatural Positions of the Lamb.

These are too numerous and varied to receive a notice here. The reader will find the subject copiously dealt with in connection with parturition in the cow in "The Cattle Doctor," and, as regards the mare, in "The Horse Doctor,"* from which general principles may be gathered.

Inversion of the Vagina

Arises in conjunction with parturition and inversion of the womb. When it is distinct it is seen as a red soft tumour protruding from the birth-passage, having a depression or opening at the centre or back end. The organ should be speedily returned, or mortification may ensue, and a truss afterwards applied to prevent a recurrence. Internally Anodynes. Dress the vagina with Antiseptic mixture No. 8, page 85.

Inversion of the Womb

Is a serious accident in whatever stage. Complete cure depends upon a speedy return and removal of original causes. Mortification is likely to arise from delay, and succeeding this death of the ewe. To return the womb, place the ewe on her back. Wash the organ carefully with tepid water to remove all foreign substances, and place a soft towel double beneath it. Smear it with Antiseptic mixture No. 8, and by gentle pressure upon each side, as well as from the thumbs at the base, the organ may be returned without much difficulty. The hind quarters being raised will greatly facilitate the operation. Afterwards the truss should be applied, or the organ may descend again. Internally, Anodynes.

Rupture of the Womb

Is known by sudden cessation of all efforts to expel the lamb, followed by collapse and death. The lamb will generally be found among the intestines.

* London: F. Warne & Co.

Vaginitis.

Inflammation of the Vagina follows parturition as a result of injuries, as bruises and lacerations. At first the discharge is pustular, and if allowed to continue it becomes thin and ichorous; it then changes to white, becomes thick, and hangs about the vulva and tail in rope-like masses. Fever always accompanies the acute stage, and may continue more or less as the disease becomes chronic. The discharge also varies as time passes, both in amount and consistency—sometimes white and abundant, at others it contains blood, pus, &c.

Treatment.—Febrifuges during systemic disturbance. Astringent lotions to the vagina, also antiseptics. Tonics as soon as the state of the pulse, &c., indicates a necessity for a change. The attendant should observe the same care with regard to this disease as it is advised for "Heaving Pains," page 139.

Urethritis,

Or Inflammation of the Urethra, in males, as a result of irritation, calculi, &c. Pain is evident in urination, and perhaps only a few drops are voided, which contain pus. There will also be swelling and discharge of pus at other times. Constitutional disturbance is present, but varies with the amount of irritation. The disease is less common in rams than in bulls, as the former do not gain speedy access to the ewes after parturition.

Treatment.—Turn the ram on his back, and draw out the penis if possible, or if swelling prevents an examination, slit open the sheath. Remove foreign substances, and inject healing fluid No. 4, page 96. Touch the ulcers with lunar caustic or the budding iron, or amputate such portions of the penis that are diseased, for which a veterinary surgeon will be required.

Dropsy of the Abdomen,

Attending or existing prior to Parturition. See "Peritonitis."—Dropsy of the abdomen often arises in ewes before and at the time of parturition, as a result of the too exclusive use of bulky succulent food, as turnips. The effusion proceeds from interference with the functions of the liver, or peritonitis in connection with parturition, but the cause is the same. The lambs are usually born dead, or if alive they survive but a few days, the abdomen being distended with pale thin fluid like that of the mother. The ewe has been saved by tapping. *Prevention* consists in allowing less turnips and more nutritious food.

Mammitis.

Inflammation of the Udder, Garget.—This is marked by pain, swelling, and suppression of the milk. The disease may be confined to one part of the gland, which is hard, knotty, and painful. It may terminate in resolution, abscess, or gangrene.

Treatment.—Aperients No. 2, 3, or 4, page 86. Draw the udder frequently, or hold the ewe for the lamb to draw it. Febrifuges No. 1 or 2, page 94. Fomentations to the udder, which should be covered afterwards by dry cloths, and if possible supported by bandages passed over the loins, &c. Apply belladonna ointment to the outside, the sedative Embrocation No. 2, page 93, or Antiseptic mixture, No. 8, page 85. Open abscesses as they maturate, and dress with healing fluid No. 4, page 96, or Antiseptic mixture No. 8. If the teats are closed, pass a pointed needle up, or open by means of the *bistouri caché*. If gangrene sets in, support the system by means of Tonics No. 1, 2, or 3, page 99, and apply Antiseptics, particularly No. 8. Treat indurations by Embrocation No. 5, page 92. Put the ewe into dry comfortable quarters away from others, especially those in parturition or that have not yet lambed, and observe the same sanitary care as advised under Heaving Pains. Preserve the udder from contact with manure, the ground, &c.

Breaking Down.

See "Wounds of the Abdomen," chapter xix.

CHAPTER XIX.

Local Injuries—Wounds of the abdomen—Breaking down—Wounds of the Bowels—Injuries to the mouth—Wounds of arteries and veins—General wounds—Fractures—Dislocations—Sprains.

Injuries to the Abdomen.

It is by no means an easy task to obtain a correct idea of the amount of damage inflicted by injuries to the walls of the abdomen. If the sufferer is not observed until some hours after the accident, the resulting swelling may greatly interfere with a proper examination. In order to ascertain not only the exact nature of the lesions, but also to be able to form an idea of the probable issue of the case, attention should be directed to the animal at an early period. Even in recent cases it is often difficult to ascertain exactly how matters stand. The muscles only may be injured; sometimes the strong fibrous membrane which forms a large portion of the floor of the abdomen. The absence of external wounds is not always a favourable sign; on the other hand, the most fatal issues have occurred when there was not a scratch to be seen, the stomach, liver, bladder, and even the heart and lungs, being seriously injured. We have also observed that many cases in which the absence of external marks, &c., indicate trivial injuries, terminate in the most unfavourable manner. For a time the sufferer exhibits to an ordinary observer an apparent freedom from pain, &c., and the appetite is scarcely affected. After a few days, however, the food is refused, and the animal becomes dull, gradually sinking from that moment, and quietly dies from an accumulation of fluid within the abdomen. Inflammation of the peritoneum has been set up, and followed by effusion; a species of internal drowning takes place, by interfering with all the organs essential to life.

Treatment.—We can only recommend such means as are indicated by external signs. When symptomatic fever is present, Febrifuges No. 1 or 2, page 94, two or three times daily. Swellings should be fomented according to directions given at page 94. If the skin has been divided, the edges may be brought together as directed under General Wounds. But before such acts of surgery are performed, the operator will make himself sure there are no particles of glass, stones, grit, sand, dirt, splinters of iron, wood, or even thorns, &c., within the wounds. Masses of coagulated blood sometimes form considerable swellings, and must be removed with care. All undue probing and interference with wounds must be avoided, as positively hurtful. A clean wound should not be washed or fomented except under professional orders. If the bowels are constipated, administer a moderate purge in the form of draught, Aperients No. 1, 2, 3, or 4, page 86. The diet should consist of articles which are light and of easy digestion. Grain, as oats, barley, or malt, swollen by hot water, with a portion of bran added, and given when cool, will be very suitable; green food, as clover, or grass, &c., and the roots are likewise needful from time to time. All dry solid food must be avoided until fever is abated and the purgative has ceased action. Movement, of course, must also be countermanded as detrimental until the pulse and respiration indicate the functions of nature are no longer disturbed. Perfect cleanliness on all hands must be insisted upon as a safeguard against blood poisoning; and plenty of fresh air should be allowed without risk of cold draughts. Lastly, the animal should be placed by himself at some distance from others.

Wounds of Bowels.

When, in addition to laceration of the walls of the abdomen, the intestines are known to be wounded, there is ample cause for alarm. Sometimes a surgical operation terminates favourably. In the meantime, before assistance arrives, the animal must be restricted from movement, in order to avoid further injury or the insinuation of foreign substances. A support to the abdomen may be needed, as well as to counteract further protrusion of the bowels. The closing and subsequent management of this class of wounds are of great delicacy, and entirely beyond the scope of amateur treatment.

Injuries to the Mouth.

The lips, gums, and checks are not infrequently injured by the edge of the drenching horn, when stupid men use it as a lever to open the mouth. The tongue is also torn or bitten when drawn outwards between the teeth. Occasionally also large and severe excoriations arise from the use of strong remedies not sufficiently diluted with water. These give rise to much swelling of the tongue, fauces, lips, and cheeks, with profuse salivation and inability to take food.

Treatment.—Astringent lotions to the mouth, No. 3, page 87, or the Electuary No. 1 or 2, page 92; also Febrifuges No. 1 or 2, page 94, when constitutional disturbance is present.

Wounds of Arteries and Veins.

These are not very common among sheep. The indications are a persistent flow of blood, the colour of that issuing from an artery being scarlet, venous blood being darker or modena red. Arterial blood is further known by the active pulsations, or spirts in the flow, the stream of venous blood being continuous or passive. To stop bleeding is not always an easy matter. In the absence of professional assistance the following methods may be adopted with advantage. First, apply pressure by means of the hand or a finger. Second, plug the wound by means of soft tow or cotton-wool, and apply pressure by means of a bandage. Third, if the wound is in a limb, apply a bandage *above* the part, if possible exciting pressure upon the main artery, which may sometimes be detected by its pulsations on the inner side. The effectual arrest of bleeding from an artery may call for an operation, such as cutting down upon the vessel at a higher point, and placing a ligature upon it.

General Wounds.

These are of several kinds: *incised*, *lacerated*, *contused*, and *punctured*.

Incised Wounds are the result of a cut, or clean and even division of the skin and tissues. They are serious, according to extent, as arteries or veins may be severed, thus producing fatal hæmorrhage before assistance can be procured. If blood is largely escaping, it should receive first attention (see Wounds of Arteries, page 162). The edges of incised wounds, being usually free from foreign substances, may be brought together by the twisted, interrupted, uninterrupted, or quilled sutures.

Lacerated Wounds are those lesions of the integument, &c., in which the edges are rather torn than cut, being irregular or ragged, as a result of blows from a blunt instrument, or falling, and tearing by hooks, horns, teeth, &c. There is seldom any danger from bleeding, but the tendency to slough is great, as the vitality of the parts is often much destroyed. All foreign substances should be removed, if possible, without washing; but, if needful, fomentations and warm poultices may be applied. Use sutures as required, or, what is better, the double many-tailed bandage, which may be glued on if the wool is shorn. This will bring the parts together very effectually. If the wool is long, this may be imitated by twisting it into ropes on each side, and tying them together.

Contused Wounds.—These are frequently little more than simple bruises. They may partake of incision and laceration also, usually being the result of falls, blows, &c. Hot fomentations continuously applied, followed by Embrocations Nos. 1 to 4, page 92, as needed by way of external applications. Scarifications to relieve local turgescence, and Febrifuges No. 1 or 2, page 94, to combat constitutional disturbance.

Punctured Wounds are frequently dangerous, as the extent of injury may be greater than signs at first may indicate. The accumulation of pus may be facilitated by closing of the original puncture, and, in consequence, it will destroy tissues and burrow beneath others in a surprising manner to a great extent. Sympathetic fever, being persistent, is a strong indication of the serious nature of the injury inflicted, and should be met by the usual febrifuges and aperients. Foreign bodies must be removed, and an outlet insured for the evacuation of pus, if possible. Dressings may be injected by means of a syringe.

Punctures of the Joints, if small, may be closed by the actual cautery, lunar caustic, or chloride of zinc; when larger, they need especial measures, which are only at the disposal of a veterinary surgeon.

The best application for open wounds is the healing fluid No. 4, page 96.

Breaking-down.

This is a formidable state in numerous instances. It consists of rupture, more or less, of the muscles as well as fibrous expansion which form the walls of the abdomen. It takes place in ewes when, in addition to their being heavy with lamb, probably having more than one, they are subjected to a bulky and not over-nutritious food, as turnips. A mass of this cold aliment robs the system of much of the necessary animal heat; thus the systems of both mother and fœtuses are reduced below par. If the ewe goes her full time, the nature of her injuries may be small, but the parturition will be difficult, and the value of the creature as a breeding animal is lost. It may happen, however, that the ewe loses condition rapidly, and becoming weak, the muscles give way, and both mother and lambs are lost, the first dying from the original causes, and the latter in an ineffectual attempt to bring them into the world. The remedy is a restricted use of turnips and substitution of lighter nutritious food.

Fractures of Bone

Consist of four kinds—*transverse*, *oblique*, *comminuted*, and *compound*. Those of the transverse kind admit of easy adjustment and subsequent rapid union. The oblique present greater difficulties in maintaining the position essential to cure; and the comminuted and the compound rarely prove amenable to treatment, owing to difficulty in keeping the sufferer still.

When fractures take place in bones of the extremities, below the stifle or elbow, an attempt should be made to promote union, as stiffness in sheep is not of much moment, and the animal may be fatted for the butcher.

The usual method of treating fractures is as follows:—First place the bones in direct position for union, and preserve it by binding on splints, by means of bandages, all hollows being filled and prominences covered with soft tow to avoid abrasion. Suitable wood splints are made by the instrument maker, and strips of stout gutta-percha may be readily adapted. The latter are first softened in warm water, moulded to the limb, and then bound upon it. Starch bandages also form suitable means for keeping the parts together, and plaster of Paris may be put on in the form of paste. These on drying become so unyielding as to prevent any movement whatever, and, when properly arranged, have been retained from first to last.

Dislocations are rare among sheep. As in cattle, the anatomical arrangements do not, as a rule, admit of laxation without serious injury to the surrounding parts. The only exception to this is *dislocation of the patella*, or knee-cap. This may arise in animals during sudden and unusual movement, when they are debilitated, also by bad management, and in some instances by becoming fast, or thrown down in awkward positions. It is likely, too, that reduction will take place spontaneously, but if it does not do so, the animal should be placed on its side, the foot pulled parallel with the abdomen, while with the other hand the bone may be pressed into position. Dislocation of this bone will cause the animal to drag the foot upon the ground, and it is possible that pain and inflammation may be present. Strong liniment or a blister should be applied after the bone is replaced.

Sprains of Tendon

Are accompanied by heat, pain, and tenderness on pressure, swelling and lameness in recent cases. Early attention is needful to avoid prolonged suffering and loss of condition. Febrifuges No. 1 or 2, page 94, and apply to the affected parts cooling lotions, as No. 1, 2, or 3, page 96.

CHAPTER XX.

Diseases of the nervous system—Phrenitis, or inflammation of the brain—Apoplexy—Epilepsy—Hydrocephalus—Paralysis—Hydro-rachitis, or louping-ill—Tetanus, or locked jaw—Rabies.

Phrenitis,

Or *Inflammation of the Brain*, is caused by falls, blows, &c., which lead to fracture of the bones of the head; sudden transitions from cold to heat; in some districts, feeding from distillery refuse; tumours on the brain; ergotised grasses, or grain so affected when mixed with the food; and plethora, especially after a time of spare or inferior keep. In some cases the animal is dull and stupid; at others it is excited to frenzy, which is, however, paroxysmal. Symptomatic fever and animal temperature are very high.

Treatment.—Strong Aperients, as No. 2 or 4, page 86, followed by a Clyster, No. 1, page 90, to a quart of which half an ounce of spirits of turpentine may be added, and repeated every half-hour. Cold water, ice, or Lotion No. 1, page 96, applied to the head. Bleeding from the jugular vein. Subsequently Febrifuges No. 2, page 94. Strong blisters, No. 1, 2, 3, or 4, upon each side of the neck. When convalescence is established vegetable tonics only should be used, and with great caution.

Apoplexy,

As a disease of the nervous system, is due to degeneration and change of structure in brain tissue, &c., chiefly arising from congestions and unnatural plethora. It is preceded by dulness and altered gait, &c., but as a rule the sheep, apparently in the best of health, suddenly drops insensible, motionless, and powerless, breathing loudly and with difficulty, at length dying very quietly. Sometimes partial consciousness is restored, and convulsions may precede death.

Treatment can only be of service before the acute signs are developed, when powerful aperients should be administered, and strong embrocations or blisters applied to the spine, loins, &c.

Epilepsy

Consists of sudden loss of sensation and all control over the natural movements, &c., of the body, associated with violent painful contractions of the muscles of the whole system known as convulsions. The animal is, at first, usually dull or stupid for some time, when in the absence of other signs he falls, bleating, to the ground, the eyeballs being fixed, jaws firmly clenched, and froth issuing from the lips. At the decline of spasm, consciousness returns, but for some time afterwards considerable dulness and languor remain. The attacks vary in accordance with the causes.

Epilepsy appears to be due to some peculiar condition of the nervous system brought about by weakness, disease, or disorder of other organs, which limits the nutrition of the brain, or interferes with its usual functions. In sheep it is due to deficient management, exposure to cold, irregular feeding, &c., worms in the nasal sinuses, stomach, or intestines, and in some cases it has followed fright, accidents, &c.

Treatment.—Remove the cause if possible. Reduce the plethoric, and strengthen the anæmic, giving iron and vegetable tonics combined.

Hydrocephalus.

Water on the Brain.—This disease is common to the lamb, which rarely lives more than a few days. The head is large, and the accumulation within gives rise to a stupid appearance, staggering gait, and tendency to move in a circle. As a rule the disease precedes birth, when the enlargement of the head is such that delivery is impossible until the contents of the skull are evacuated. The disease is of a scrofulous nature.

Paralysis.

Palsy, or loss of power, rarely occurs as a distinct affection. It is usually estimated a sign of other diseases,

being a termination nearly of all nervous as well as many other affections. Lesions of the spinal cord, resulting from falls, &c., are the usual causes of paralysis as a separate disease. A cure is hardly worth attempting, except under very special circumstances.

Hydro-Rachitis,

Or *Leaping-ill*, *Trembling*.—This peculiar disease consists of peculiar nervous action which sometimes draws the neck or body into strange contortions, with which are associated stupor, trembling, &c. In many instances the lambs are paralysed at birth, sometimes behind or on one side, at others the fore legs only are affected. If free at birth, paralysis follows the attack of muscular spasm which we have referred to. As the animal walks the action is peculiar, a dropping of the limbs and jerking of the body being constant. From this circumstance the idea of leaping is conceived, hence the term "leaping or louping ill." From the fact that swallowing is sometimes difficult or impossible, the malady has been called "thwartil ill," which we are inclined to construe as "throttle ill." It is, however, a disease of the spinal cord, and has been hitherto supposed to be common only to the north of the Tweed. We have, however, met with it in the county of Buckingham. Lambs and sheep under one and a half years old are the common victims. The cause of the malady has hitherto been one of conjecture. Close attention has, however, developed the strong presumption that it is to be found in ergotism, derived from a diseased condition of seeding grasses.

Treatment.—As for paralysis generally. It is, however, rarely of any avail. Prevention should be sought by removing lambs from land which is known to produce the disease, and afterwards institute such changes as will insure its being more uniformly productive.

Tetanus,

Or *Locked Jaw.*—A violent irritation of the nervous system, which is manifested in a continual spasm or contraction of the voluntary muscles of the body, those of the jaws being implicated, and firmly closing the mouth, hence the name "locked jaw." It arises from cold and exposure, wounds and injuries, &c., castration, and, occasionally, docking; sometimes, without any obvious cause. It is said to occur in lambs when ewes are overfed with trefoil.

Treatment.—Strong Aperients No. 2 or 4, page 86. Feed on bran principally, held for the animal in a suitable vessel. Absolute quiet is indispensable. The remedies must be supplied by a veterinary surgeon.

Rabies.

This formidable affection is, unfortunately, too common, being due to the bite of rabid dogs. The signs and progress of the disease have been entered into at some length in "The Cattle Doctor." An intolerable itching always takes place in the situation of the wound first inflicted. Although it has healed, to all appearances perfectly, yet the violence of the animal is such as to tear it open in some cases. The desire for water is very great, but there is no power to swallow. These may

subside, and the sheep becomes stupid, standing with head erect and to one side. At other times, there may be excitement, with diarrhœa, profuse salivation, mournful but constant bleating, fright from the rustling movement of persons or objects, as paper, with occasional attempts to butt. These give way to depression, weakness, and exhaustion, and finally paralysis and death.

CHAPTER XXI.

Diseases due to parasites—*Œstrus Ovis*, or grub in the nasal sinuses—Hoose, husk, or verminous bronchitis—Gid, or hydatid disease of the brain—Measles—Diseases of the liver due to parasites—Hydatids—Echinococcus parasitism—The fluke disease, or liver rot—Worms in the digestive canal—Scab—Lice, ticks, and maggots.

THE diseases which are due to the presence of parasites are being more definitely understood each year. The list given above by no means represents the extent to which an enumeration might be carried. It will be sufficient for the amateur. The close congregation of animals is highly favourable to the propagation of diseases of all kinds, the transmission of parasites especially, and as a fertile cause of serious mortality; and, as far as mankind is concerned, he is reminded that sheep, no less than other animals, harbour in their bodies a number of these unwelcome guests, kindly accommodating them in their stagal development, until the time arrives when they may be transferred to the human body in the flesh used as food.

Œstrus Ovis,

Or *Grub in the Nasal Sinuses.*—The gad-fly, which torments bovine animals and deposits its eggs in the skin of their backs, is represented by a smaller variety, which prefers to attack the sheep. For the purpose of propagating its species, it alights on the nostrils and deposits its larvæ there, and this, by means of its wormlike motion and the assistance of its hooklets, finds its way upwards to the sinuses or the back of the nasal passages. There it remains, securely sheltered, throughout the winter and following spring, gradually going through a higher development, until the warm summer weather induces it to descend to the ground. The rays of the sun now speedily complete the necessary changes. It mounts on wing, and devotes an ephemeral existence in conveying its larvæ to the nostrils of the sheep, and finally dies.

Irritation does not necessarily arise from the presence of the parasite within the sinus, &c., but there are instances when pain and inconvenience are extreme. Inflammation of the mucous membrane succeeds, with excessive pustular discharge, probably from one nostril only, dulness of the sinus on percussion, irritative fever, &c. There may be also stupor or dulness of spirits, and even epilepsy. The sheep when attacked sneezes, stamps with the feet, and violently rubs the nose against the nearest object or in the soil.

Treatment.—If the parasite is to be removed, the sinus must be opened by means of a trephine, and astringent lotions applied will act favourably.

To prevent the attack of the gad-fly, smear the outside of the nostrils with wood tar, or "Sanitas" ointment, &c.

Hoose, or Husk.

Verminous Bronchitis.—This is a troublesome disease of lambs, which in some years prevails as an enzoötic, with all the virulence of a plague. It consists of a troublesome and constant cough, due to the presence of bundles of worms, which are coiled up in the mucus of the bronchial tubes and air cells. In some seasons older animals as well as lambs exhibit the effects of the worms located in the intestines, creating violent diarrhœa and dyspepsia. This causes them to devour the soil or sand in immense quantities, as well as drink water to the danger of bursting.

Treatment.—In our present experience of this disease it is found that internal remedies have but a very limited effect upon the parasite. When out of the body it will submit to drying for weeks, to the action of strong acids and chemical solutions, and yet preserve its vitality. How, then, can we expect good from medicines unless they are so strong as to kill the animal itself? for it is clear that must be done if we attempt to kill the worms. The great principle seems to be that of fortifying the system against the ravages of the parasite by administering tonics. Turpentine and lime-water, it is found, have a certain beneficial effect. The more recent method consists of injecting special remedies direct into the wind-pipe, for which, a syringe, armed with a hollow needle, is required. With care the results are beneficial. Fumigations with burning sulphur are useful, likewise the vapour of turpentine and chlorine gas. The safe way of making use of fumigations by sulphurous acid and chlorine gas, is to drive the animals into a building having doors and opening windows on several sides. The operator should enter with the animals, and, when all are closed, generate so much gas as he can inhale without distress for some time. At a given signal his assistants may open the doors and windows to admit fresh air. This process should be repeated daily, or twice if needful.

The *prevention* of this, in common with all other diseases due to the presence of parasites, is a subject of the greatest importance to the general stock-breeder. There is, doubtless, much to be done by way of improvement of pasture land. We are of the opinion that very old swards need putting under the plough, and the use of common salt and lime would be beneficial, the turf being pared and burned. This is of vast importance when preserves are near, and winged game, rabbits, &c., abound. Lastly, the lungs and intestines of all affected sheep should be *burned—never buried.*

Gid, or Hydatid Disease of the Brain.

Sturdy or Turnsick, Gid, Goggles, Turnside, Vertigo, Hydrocephalus, Hydatidæus.—This form of parasitic disease is developed in the brain. A cyst or bladder, in its gradual formation, interferes with the structure and functions of the organ, giving rise to peculiar movements, from which the above terms, with others in common use, have been derived. Sometimes the sheep continually moves in a circle, to the right or left according to the side or hemisphere affected. In other instances the head is held high, and the feet are raised unusually high also as he goes straight forward; this indicates the hydatid is situated between the hemispheres. When the lesser brain or cerebellum is the seat, the sheep has no control over his movements; he therefore twists, falls, and rolls about in awkward helplessness. The cause is due to the ova of tapeworms, scattered by dogs on the pastures, being taken into the stomach, and carried it is supposed by the circulation, they develop in the situations referred to. The dog, in turn, obtains the tapeworm by devouring the hydatid, when the sheep's head is given to him for a meal.

The presence of the cyst in the brain is known by softening of the bones of the skull over the immediate locality. This spot is punctured by a trocar, to which a syringe is applied, the up stroke of which exhausts the cavity by withdrawing the cyst. The disease often proves fatal on account of the amount of structural derangement which may have taken place within the brain.

Prevention consists of avoiding the use of dogs for collecting and driving the sheep; or, where they cannot be dispensed with, to destroy the heads of the affected sheep by burning instead of allowing the dogs to consume them. The latter should also be specially treated for tapeworm, the fæces containing either worms or ova being destroyed by fire.

Measles.

On the authority of Dr. Cobbold, another variety of tapeworm—*Tænia tenella*—common to man, depends upon the sheep for its last developmental stage. The ova of the worms are doubtless transferred to pastures, &c., by means of manures, the herbage of which is consumed by sheep, and thus they gain access to the stomach. At a later period they migrate to the muscles, where they become encysted, and are incapable of further harm to the host. Mutton, half cooked, conveys the cysts again to mankind, and in his digestive organs the fully developed worms are produced, often creating serious disturbance. During the life of the sheep there are no visible signs by which the migrations of the cystic worm can be definitely recognised, unless we attribute the various forms of stiffness, rheumatism, &c., to them. Thorough cooking of the meat destroys the parasite.

Parasitic Disease of the Liver.

The liver of the sheep is liable to attack from several parasites. The first to which we shall allude is the *Cysticercus tennuicollis*, which appears in the substance of the liver as a cyst, or similar form to the hydatid of the brain, causing the disease called "sturdy" or "gid." This also is one of the stagal forms of a variety of tapeworm, *T. marginata*, and the dog is the animal from which the sheep derives the ova. It is believed that three dogs out of every twelve are infested with this tapeworm, and the announcement should put our farmers on their guard in order to suppress the spread of the ova, by not allowing dogs to eat at any time the liver and offal of sheep slaughtered and dying of various diseases, except they are very highly boiled. Such organs, when known to be affected with hydatids, should always be destroyed by burning. The disturbance set up in the liver by the cysts, will, of course, partake of the usual characters of liver disorder generally.

Another parasite, and by far the most remarkable of all, is the *Tænia Echinococcus* of the dog, the larval form of which obtains access to the liver of sheep, &c. The fully developed worm does not grow to a larger size than probably one-third of an inch, but when the larvæ reach the bodies of animals, and even man, they assume such dimensions that one of these is estimated to be many thousand times larger than its adult form.

The subject of echinococcus parasitism, to say nothing of any other form, when attentively studied, fills the mind with nothing less than alarm. The system of agriculture is materially active in aiding their propagation, and the question of public health demands that our live stock should be placed under more direct veterinary supervision.

In this, as in other forms of parasitic disease, the ailment will partake of the manifestations peculiar to the diseases of the organ infested, as congestion, inflammation, &c. It is not always possible to pronounce

decidedly during life that such states are the result of parasitic invasion; death only reveals the condition. As our acquaintance is enlarged, we shall be able to trace the migrations, and probably their presence, more accurately than at present.

With regard to *prevention*, we must refer the reader to the advice given under the preceding, as well as other forms of parasitism.

The third form of parasitic disease of the liver is termed—

The Fluke Disease,

Or *Sheep-rot.*—This is another of the most frightful diseases to which our flocks are liable, of which the seeds are laid during the warm weather of May and June. Sheep have been known, by a single access to a marshy field at this time of the year, to have become affected with the rot, which has proved fatal during the following autumnal and winter months. The disease was extremely fatal during the winter season of 1860,*

* Previous outbreaks are recorded as having taken place in 1735—6, 1747, 1766, 1776, 1792, 1810, 1816—17, 1824, 1830—31, 1833, 1853—4. Lastly we have the fearful visitation of 1880.

and unusual attention was then directed to it. Professor Brown, of the Cirencester College, contributed a very instructive paper on the subject to the pages of the ninth volume of the Bath and West of England Agricultural Society, and Professor Simonds gave a most instructive lecture on it before the English Agricultural Society. From the latter we take the following extracts:—

"The now generally received theory of the disease is founded entirely upon the existence of the fact that, during it, we find certain entozoa (flukes) inhabiting the biliary ducts of the liver. These entozoa produce a number of eggs; these eggs pass out of the liver into the intestines, and are consequently expelled with the feculent matter of the sheep in countless myriads. The original form of the theory was that healthy sheep, if put upon pasture grounds where these eggs exist upon the soil, receive them into their organisms; that the egg produces the fluke, and that the fluke would consequently find its proper *habitat*—that it would seek out instinctively the biliary ducts of the liver, where it would locate itself and grow to its own perfection and the ultimate destruction of the sheep; but it is found that fluke eggs do not immediately produce flukes. Some ten years ago I put this to the test of positive experiment. I collected a great number of the eggs of the fluke—far more than it would be possible for a sheep to receive into its stomach in the course even of a summer's grazing. I took not less than a teaspoonful of them, and it would be scarcely possible for you to count the number in a single drop of the water in which they are placed under the field of the microscope; and these, to the number of millions, I conveyed into the system of a sheep, which I kept six months, and then had it destroyed. On examining its liver and other organs, I found that there were no entozoa at all in the biliary ducts. In reality, there was not a single fluke produced from those millions of eggs so carried into the system of the sheep.

"The fluke, or the *Distoma hepaticum*, as it is called, is so designated from having apparently two mouths or sucking discs—one placed at the anterior part of the body, which may be truly regarded as a mouth; and the other placed at a short distance below the neck, just where it terminates in the body. The animal inhabits the biliary ducts of the diseased liver; and if we slit up these ducts in any case of this kind, we shall find that they are filled more or less to repletion with these *distomæ*. They ultimately lead to an anæmiated condition of the entire organism. They not only feed upon the bile which is produced, but they alter the structure of the liver by their presence, just as we find with entozoa in other parts; and when this is the case, it is of course perfectly impossible for the liver to secrete healthy bile, any more than pure water can flow from an impure spring. The bile plays an important part in the manufacture of blood, and if it is not in a healthy condition, pure blood cannot be produced from the food which the animal takes. The result is, that the entire organism is supplied with impure blood, while the system is being drained by the presence of these creatures. After a certain length of time we find that dropsical effusions take place, and we have the disease established in all its intensity and in all its destructiveness.

"It is not exclusively a sheep disease. These creatures have been found not only in cattle but in pigs, in the ass, and also in the human subject. They are very widely dispersed, but it is in the sheep in particular that they accumulate in such numbers as absolutely to produce this specific malady.

"There is one circumstance especially which renders the ova of the fluke especially interesting to us, namely, that if we examine them never so carefully, and any number we please, those that have been naturally expelled from the creatures as well as those that are contained within them, we shall find that there never exists within the ovum anything of the outline of the young fluke; and this fact being established, it is evident that in order for the fluke egg to produce ultimately a fluke, the germs contained within it must pass through a series of transmutations; and it is by studying these that we can get at something like valuable information with reference to the manner in which these creatures are propagated. It is now several years since I thoroughly convinced myself of that fact, and I believe nearly every person who is at all observing or familiar with the circumstance knows that if you slit up the gall ducts of sheep or any other animal affected with distoma, though you may see some of these creatures smaller than others, you never see what might be called a number of young flukes; nor will you be able to discover a young fluke by microscopically examining the bile in which they live. This, then, together with the other circumstance to which I have referred, shows at once there is not a reproduction of the entozoa within the biliary ducts; so that (to put it in a practical way), supposing a sheep to receive six of these flukes into the biliary ducts, they would never multiply. They would deposit millions of eggs in the smallest ramifications of the ducts—instinctively deposited there, it would seem, for the purpose of preventing them from too readily flowing out by the functions of those organs—but you would never have more than six flukes. Now, what does this explain? It explains a fact of every-day occurrence. A person will often tell you, 'I sent a lot of sheep to the butcher; I never had more perfect and beautiful animals; they were as fat as sheep could be; yet the butcher found eight or nine, or ten or twelve flukes in the liver.' The fact is, that here they did not exist in sufficient numbers to lay the foundation for disease. If it were otherwise, we can see that if one fluke only passed into the biliary duct, it would multiply almost *ad infinitum*, and the animal must fall a victim to the affection induced thereby. Here, then, we have a practical result at once, arising with many others from an investigation of the natural history of this creature.

"Now, what I am going to explain will only allow of an analogy with regard to these creatures. We believe that each of these fluke eggs contains a number of moving ciliated cells, which are more or less round; that these creatures passing into water, for example, are set at liberty; that they become parasitic on some of the creatures which are met with in water, and that, when so parasitic, they have the capability of propagating themselves; that they subsequently pass through a series of changes, and again become, as we believe on analogical grounds, parasitic a second time on other creatures, when they change into the fluke-like form. So that, tracing the process all through, we should say that the egg sends forth a circular germ which is ciliated, and which has a rotary motion in water; that in this condition it becomes a parasitic on molluscs, small snails, and things of that description; that when it gets into the body of these creatures, working its way in just below the skin by the cilia which it possesses, it undergoes a perfect change, becoming something like a chrysalis, in which condition it propagates itself; that the creatures immediately coming from it reach a certain order of development in the snail; that they then escape from it in the form of the so-called cercaria, swim about in water, and after a certain time again become encased, and again become parasitic on the snail, in order, in reality, that they may reach a higher form of development; they then remain in the water, attached either to plants or to smaller creatures inhabiting water, especially slugs and snails; and when these things pass into the stomach of the sheep, they find their proper *habitat.* We now get them developed into flukes; and it is important to bear in mind that the last transformation takes place, not in the liver, not in the biliary ducts, but in the stomach of the animal. You will see presently what importance there is attaching to that fact, which has only very recently come to light.

"Now, if we accept this as an approximation of the truth with regard to the development of these things, we shall find that it unravels the whole mystery with regard to rot.

"It explains at once why certain districts are dangerous. It explains to us the occurrence which has been recorded over and over again, that out of one hundred sheep, for example, ninety-nine have strayed over a common, and that one has been accidentally prevented from doing so; that subsequently the ninety-nine have been attacked with rot, while the one that remained behind escaped.

I now believe that it is perfectly possible for sheep to be free from the cause of rot at this minute, and to receive it at the next; that is, if they are placed under circumstances where they can obtain these creatures in one of their forms of development; and they must have reached a particular stage of development before they so receive them. Now let us suppose we have the cercaria rolled up into the form of a creature something like a chrysalis covered over by its shell; let us suppose that which is positively the fact, that hundreds of these creatures can be seen upon small molluscs, many of them also loose in water, that they are adhering likewise to plants and growing in damp situations; then a sheep has only to drink a mouthful of water to take in an indefinite number of these little things, and, taking them into the stomach, they there become developed into flukes. We have, therefore, very good ground for believing that rot can be very readily and rapidly received. People want to know how it is that we get rot in certain seasons and not in others. Whenever there is an excess of moisture, certain pastures, which are perfectly free from disease at other times, become affected. This is easily enough explained. We have seen that the ova of these creatures come out in millions from one affected sheep: what, then, must be the number of them in hundreds of sheep so affected? We know not what is the duration of the life of these ova. It is possible that they may remain years without undergoing change, until placed under favourable circumstances to undergo that change. If, then, we look at the vast numbers in which these ova exist, and at their power of maintaining their vitality for a great length of time, we get rid of a considerable amount of difficulty. Then we find that when you get an excess of moisture, and with that an elevated temperature, we have the living germs, that are separated by the bursting of the egg, set at liberty, and becoming parasitic on other creatures, as I previously explained, passing through that series of transformations comparatively quickly, and being very easily received into the system of the sheep. It is notorious that sheep placed upon watery meadows receive the rot;

but it is equally notorious that it is only at a certain period of the year that they will do so. You may, as every practical man knows, put sheep upon water meadows during the winter months, or in the early part of the spring; but if you water your meadows in the month of May, and then get a luxuriant herbage springing up afterwards, and put the sheep on this, you are almost certain to rot them. As you approach midsummer the danger increases; and as you approach towards winter it decreases. That single circumstance shows that, when there is moisture and heat combined, the cause is brought into operation. I have spoken of the great losses amongst sheep affected with rot. When were the sheep affected with rot that we are now losing? I answer, last midsummer. That is the time to which you have to look. The cause was received then; the development has been going on in these creatures since that time. They have now attained their full size, and they are producing mischief in anæmiating the animals.

"With regard to the symptoms by which we recognise rot, it is universally admitted, I believe, at least by all practical men, that in the earliest stage there is no great depreciation of the value of the sheep, that it does not apparently suffer any inconvenience, but that it rather accumulates flesh faster. The explanation of this is easy enough. The small fluke enters into the liver, not perfected, but having of course to be perfected, and to have its generative system fully developed in the gall ducts of the liver. When they first pass in, although, perhaps, in very considerable numbers, they simply act as a sort of stimulus to the action of the liver; they consequently call forth an increased secretion of bile, and as there is no alteration in the character of the bile, the sheep now being fairly supplied with an ordinary amount of food, will make relatively a larger quantity of blood out of that food. The liver being in a state of excitation scarcely bordering upon disease, the sheep will lay on a larger quantity of flesh. It is very difficult, therefore, for us to say what are really the early symptoms of rot, if we except this accumulation of flesh. There can be no doubt that in an advanced stage of the disease we take cognisance of the affection very readily; but unfortunately it happens that the symptoms which then show themselves are such as to prove to us that the system is breaking up, and that the time is past for curative measures. At first, I say, the symptoms are exceedingly insidious, but after a certain length of time we find that such animals have an occasional cough, that their appetite is somewhat impaired and fastidious—to-day feeding pretty well, to-morrow scarcely at all. They will be easily acted upon by all external causes, and if exposed to wet and cold will suffer a great deal of inconvenience.

"Subsequently there is a gradual wasting of the body, and this takes place even before other symptoms which are looked upon as unmistakable proofs of the existence of the disease show themselves, at any rate, to any great extent. It is always a suspicious circumstance if you find that animals have been remarkably well, and then towards the latter period of the year, when they should be going on maintaining their condition, they begin to waste. If you put your hand upon them and find them 'lean on the back,' as it is called, that the vertebræ are sticking up and are bare of flesh, if you find animals 'razor-backed,' if I may use the expression, it is a pretty good indication that they are affected. This state of wasting being once established, it continues, and we then get a pale state of the skin, which becomes of a yellowish tint, and very frequently it may be said that jaundice to some extent becomes associated with dropsy. If you open a sheep which is in a somewhat advanced stage of the affection, fat is particularly yellow. We afterwards find that the inner angle of the eye becomes exceedingly pale, so that when we invert [the eyelids] and press forward the *membrana nictitans* of the eye, we find, instead of its being in a healthy condition, that it is covered with a number of red lines, marking blood-vessels through which the blood flows, but the blood being deprived of its red cells, the liquid is colourless, generally speaking. This evidently arises from the circumstance that there is a great drain going on upon the system, that the blood itself is being deprived of its watery matter, and that the watery matter must be taken up into the organism to make up for the drain. Then we find, further, that as the disease advances, we get a variable state of fæces: sometimes they will be scouring, and at other times nothing of the sort will be observed; and it is always a suspicious circumstance to find sheep in the autumnal months *occasionally* scouring. Of course such animals lose their strength very quickly. They are dull, dispirited, and often found lying down. As the disease advances, the breathing becomes somewhat difficult; the wool is easily removed; œdematous swellings, as they are called, begin to show themselves and to accumulate in different parts of the body, more particularly under the lower jaw. These are nothing more nor less than dropsical effusions: the blood is almost changed into water, and then you have these swellings finding their way into the areolar tissue in different parts of the body, passing freely through that, and accumulating particularly under the lower jaw, because the animal frequently has its head pendent, and in the act of feeding a gravitation of fluid takes place.

"The treatment, then, of rot, speaking of it as a curative treatment, must have for its end and object the removal of the cause; and, if we have these entozoa, which are the proximate cause of the affection, all our efforts must be made to the displacing of them from the biliary ducts; though I believe, if we take sheep in an advanced stage of this disease, and could destroy or remove from the biliary ducts every fluke contained in them, we should not save the life of the sheep, because the presence of these creatures produces mischief in two ways: it leads to organic changes in the structure of the liver, and the breaking up of the entire organism of the animal. We cannot put a new liver into the animal, and the powers of life cannot be supported by us long enough to bring it into a new state. But it is equally true that we must look to the causes, for the purpose of getting rid of the effects. There is no multiplication of these creatures in the biliary ducts. If, therefore, we were to adopt the treatment early in the case, when there are but few of these entozoa, it would necessarily be attended with success. We are to measure the danger of the animal by the number of flukes existing in the biliary ducts, and by the length of time that they have so existed. It is proved by daily experience that we can resist to some considerable extent the inroad of the disease, and we have, therefore, to look to the means which are at our disposal for the purpose of keeping the animal body together, if I may so express it, so that the ultimate loss shall not be very great. Now, how is this to be done? It is to be done, in the first place, wherever it is practicable (but it is not always so), by protecting sheep from the inclemency of the weather; and, in the next place, by abstaining as much as possible from all succulent vegetable food, all food which has an excessive moisture. We should give the animals as much nitrogenous food as we possibly can, so as to lay the foundation for pure healthy blood; and we want, at the same time, to throw tonics into the system, with a view of medicinally strengthening it. What should these consist of? It is rather a difficult thing, when a man has 500 or 600 sheep, to be giving them tonic agents, and therefore we are obliged to choose something which will of itself be unobjectionable to the animal. Sulphate of iron is an excellent tonic for purposes of this kind. Not only is it a good invigorator of the system, but an agent which sheep will take readily; but it is to be borne in mind that it is, in itself, a very great anthelmintic, and may do much good in that way. So that if, in December or January, I had taken a number of sheep affected with this disease, and the subjects of structural change in the liver, and given them nitrogenous food, protecting them as much as possible from the weather, and giving them sulphate of iron; if I had husbanded the animals' powers to the greatest possible extent, and added also, from time to time, some salt to their food, I have no doubt I should have kept them alive, and been able to sell them as fair food, at a small cost, in the market.

"I once purchased a lot of rotten sheep; I gave them no physic of any kind, but merely kept them in sheds during the winter-time, fed them with corn and cake, giving the most generous diet I could; and I not only prevented the further progress of the disease in several of these cases, but I even made the animals accumulate flesh, and they went into the market in the following spring, forming pretty fair meat for the people. This shows what can be done by generous diet and a protection of the animals. When we have animals in this condition, there will be a great advantage arising from the employment of diffusible stimulants, and such as, to some extent, are more powerful anthelmintics. For example, we may use turpentine (which the animals must, of course, be dosed with) in conjunction with sulphuric ether as an invigorating agent, and at the same time an anthelmintic. If I were to take half-a-dozen sheep, and simply give them sulphuric ether with oil of turpentine day by day, attending to those other things that I have mentioned, not neglecting salt as a stimulant to the digestive organs, I think it very likely that two or three of them (according to the stage of the disease) would be benefited by treatment of that kind.

"It is well known that sheep do not rot on salt marshes; no matter how wet they may be, no rot takes place on them. Now, what is the explanation of that fact? These infusoriæ that we have been speaking so much of are creatures belonging to fresh water, and not to salt water. If, therefore, we were enabled to take a quantity of salt sufficient to render all these damp swampy places in our meadows sufficiently salt, we should destroy the whole of these creatures, and so get rid of the cause of this affection. But we cannot do that; you could not have a sufficient quantity necessary for the purpose without destroying the whole vegetation upon the meadows. But bearing in mind the fact that the last change into the fluke takes place in the stomach, and not in the liver, and that salt is destructive to these infusoriæ, as we may call them, you will see that if you convey salt in sufficient quantity into the stomach, you may destroy them there. before they undergo the last change. They undergo their last change, as I have said, in the stomach, and then as flukes find their way into the biliary ducts; if you can prevent that last metamorphosis, you get rid of the cause. I believe, therefore, that rot may be prevented to a very considerable extent by the use of salt. When is the dangerous period to sheep? As you approach towards midsummer. We must prevent the disease then, if we can; that is the time to strike at its root. Now, what have we had in the past season (1860)? We have had a very wet summer. I happen to have been an unfortunate farmer—unfortunate, I say, because, like many others, I have not had a very profitable return this year; and I had a number of sheep, and foresaw what was coming. I said to some of my neighbours, 'We shall have a great deal of rot this year;' and I thought I would attempt, if I could, so far as my own sheep were concerned, to save them. What did I do? The sheep were on wet meadows up to the fetlock-joints nearly every day, and nobody could avoid it. But at midsummer I began to feed the lambs and sheep with corn and nitrogenized food, giving them with every meal a small quantity of salt. I continued that plan during the autumn, and I have the satisfaction of saying that I do not believe at the present time I have one of those lambs affected by rot. I kept killing them week by week to watch their progress. And here I may incidentally observe that long-continued wet weather is very prejudicial to the sheep in another way. I refer now to the water-rot. What was the state of the liver of these animals at midsummer? There were no flukes or anything of that kind, but the liver was streaked with white here and there, and generally pallid. That was for the want of nitrogenized matter. The bile cells were blanched; the liver had become structurally diseased, and it was a good *nidus* for these entozoa to inhabit. Not only, however, did the treatment prevent the entozoa, but it brought about a healthy state of liver, for in the course of a month or two I found that that organ resumed its natural colour and consistence. I again say, that if we commence at midsummer, and continue the treatment through the dangerous period of a wet season, we may do a great deal in the prevention of the disease. And I may go further, and say that even on farms where we have what are called rotten pastures, on which sheep are placed, they might be preserved to a very considerable extent simply by giving nitrogenized food and salt, to destroy these creatures within the stomach, and prevent their final change; alternating with the salt a tonic invigorating agent, such as sulphate of iron. I do not depend upon the salt alone; far from it; but it is a valuable agent, and its value depends more upon putting these things into salt water, as it were, in the stomach, than anything else. This is the course I recommend. You have to look to the condition of the liver in a wet season; you have to look to the necessity of laying the foundation for a good quality of blood, by giving these animals nitrogenized food, and throwing sulphate of iron into the organism. Every practical pathologist, human or veterinary, knows very well that if you have an anæmiated or bloodless state of the system—if there is a deficiency of the red cells, upon which the invigorating properties of the blood depend, those cells will rapidly multiply, and the blood regain its proper colour, by the use of iron. This is the reason why sulphate of iron should be employed. It should be given in fine powder, and in doses of about ½ drm. a day; not, however, that a larger quantity would be prejudicial. The sheep should be divided into small lots; and if you have about a score feeding in one trough, there should be 10 drms. of sulphate of iron mixed with the food for the day; and then, if one should get a little more, and another not quite so much, it will be of very little importance. These are matters of detail which of course every individual farmer must carry out for himself; but if he will adopt these leading principles that I have attempted to lay down in a very imperfect manner, he will save a considerable number of his sheep from falling a sacrifice to the affection which is commonly designated 'rot.'"

No apology is offered for occupying so many pages with valuable information on the sheep-rot—a subject which ought to be borne in mind by the farmer in the month of June. We have, with many others, to thank Professor Simonds and the Agricultural Society of England for placing it at our disposal.

It only needs to be added, that the first preventive of all consists in an alteration of the condition of the land. Sheep do not rot in dry pastures; and land drainage is the great remedy on which the farmer should depend for his safety. This is the essential process by which the later migrations of the embryo fluke are to be interrupted. As they are at this period parasitic in various kinds of snails, by the removal of standing pools, &c., by drainage of course we carry away also the snails. Neither can they crawl upon dry substances to any extent. This fact points out the importance of keeping sheep off the pastures until the sun has dissipated the dew and moisture after rains, &c., the snails having left the surface for cool and damp situations.

Worms in the Digestive Canal.

There are many kinds of worms which infest the stomach and intestines of sheep. We can only refer to them briefly; a minute account would fill a large volume, and even a partial one is sufficient to weary the reader.

The usual and principal varieties are round, tape, and thread worms, and these sometimes abound in such large numbers that seem almost incredible. We have frequently taken from a single sheep as many tapeworms as filled a pint measure, and others, as the round and thread worms, prevail also at times in like manner.

The only distinct proof of the existence of worms during life is the discovery of the worms or their eggs in the excrement, but in addition we may notice a general want of condition. The wool is harsh and dry, the skin dry, dirty, and irritable, from the accumulation of scurf, &c., and is tight on the body. The sheep rubs his nose and licks the floor, or swallows dirt, sand, &c. The appetite is variable, bowels irregular, with occasional diarrhœa, tympanitis, and itching at the anus.

Treatment.—Nothing answers in the hands of the sheep-owner so well as turpentine blended with linseed oil—six ounces of oil having half an ounce of turpentine, which is a dose for a large sheep. When any particular parasite prevails, special remedies in addition may be employed by the veterinary surgeon.

Scab, or Mange.

This is another form of protracted and troublesome disease, which but for the negligence and carelessness of unprincipled men—low dealers, who drive their affected sheep by stealth—might be eradicated summarily. It is due to the burrowing in the skin of an insect called a mange mite, or acarus. This operation is pursued to fulfil its natural functions, viz. obtaining food and propagating its species. Having only an ephemeral existence, the insect is exceedingly industrious; its whole life is occupied in sustaining the body and placing its eggs in a position suitable for being hatched. Thus one crop after another is raised and fresh parts of the skin are invaded. The resulting irritation is severe, and the sheep violently rubs himself against every object, tearing off his wool, producing raw surfaces, and leaving the ova of the acari upon everything he has touched. As the migrations proceed the raw places are covered by a brown incrustation or scab, hence the name; this dries and falls off, and eventually the process of healing is completed. The propagation of the insect, however, still continues, and if time is allowed and the life of the sheep is preserved, the old spots would be revisited and again subjected to the same irritation.

Treatment.—The first proceeding should be that of ascertaining whether acari are present; this is generally taken for granted, on the evidences of so much severe irritation and the rapid spread throughout the flock. The farmer adopts, as a means of cure, one of three processes, viz. *dipping*, *pouring*, and *salving*.

The process of *dipping* is by far the most effective and profitable for the cure of scab. The various materials or preparations, whose name is legion, are usually more or less soluble, or at least miscible with water, and their activity is also increased by making the solution warm. The vessel or dipping-tub is sufficiently capacious for the largest sheep, and is supplied from time to time with the mixture. Into this the sheep is placed, usually on his back, while two men, holding him down, use the free hand for working the mixture into the diseased places. After being thus immersed and acted upon for the space of several minutes, the sheep is raised, transferred to his feet, the wool pressed, and he is allowed to walk into a drainer, where much more of the fluid leaves the fleece and returns to the bath. Scabby sheep will sometimes require a second and even third dipping.

Pouring is a much more tedious process, and consumes a double if not greater quantity of the actual mixture. The process consists of parting the wool down the middle of the neck and back, when, by means of a kettle or other vessel having a spout, the fluid is poured upon the parting, a man upon each side of the animal carefully rubbing it in as it saturates the wool.

A third method, that of *smearing* or *salving*, is sometimes adopted, but as a rule it is used as a preventive of the attack. The wool is parted as just described, and the process is carried also over the shoulders, hips, and down the sides. Into these furrows mercurial ointment is smeared by the shepherd. The practice is not so advantageous as either dipping or pouring, besides it is dangerous to the animal, and the poison of mercury is not easily got rid of. The flesh is, therefore, less suitable for food.

Various preparations for the cure of scab are given at page 43. We are informed Professor Tuson's arsenical wash is a very efficient preparation. It may be obtained of Messrs. Willows, Francis & Butler, 101 High Holborn, W.C.

Lice, Ticks, and Maggots.

A very useful application may be made as follows:—Olive oil, 2 pints; spirits of turpentine, 4 ozs.; potash, ¼ oz.; hot water, 4 ozs. Dissolve the potash in the water. Mix the oil and turpentine together thoroughly, then add the potash solution and shake well.

CHAPTER XXII.

Poisons—Empirical poisoning—Accidental poisoning—Wilful or malicious poisoning.

THE subject of poisoning among the lower animals forms one of the most important branches in the study of veterinary medicine. Death arises more frequently than is known or supposed from the careless or ignorant use of remedies, and the fact calls for greater vigilance from those concerned in the welfare of stock.*

The death of animals from poisons takes place in three ways:—

In *Empirical Practice* farriers, cowmen, shepherds, and

* See the author's little manual, "The Veterinarian's Pocket Remembrancer," which contains a mass of useful information on this and kindred subjects. Published by J. & A. Churchill, London, Price Three Shillings.

many proprietors make use of remedies, the action of which they are totally ignorant; they continue the use of a remedy without being able to distinguish the fact that it is altogether unsuited to the animal. We are inclined to believe that our experience in the effects of poisons generally upon various kinds of stock has been gained more by seeing the results of empirical practice than in any other way. We can produce records of many instances of animals under the treatment of such persons as have been named above, which show that the simple administration of remedies for the cure of disease has been absolute poison. To many such cases we have been called, and only by timely withdrawal of the agent a life has been saved from sacrifice. Even the most useful and simple remedies may prove poisonous when given at improper times. Thus common salt is a valuable condiment under certain conditions, and does valuable service in rot, low condition, &c.; but at a later stage, especially in pregnant animals, it is a deadly agent. The same may be said of many other remedies. All medicines should, therefore, be used only under proper directions, or by the advice of a veterinary surgeon.

Accidental Poisoning occurs in a variety of ways, chiefly through absolute carelessness on the part of those who should have strict custody of remedies. They are sometimes carelessly thrown about, and thus become mixed with food; external preparations are substituted for those intended for internal use. Vermin poisons reach the meal-tub, and finally the sheep trough; too large doses of powerful medicines are sometimes given at too short intervals; and so the category of fatalities is made up. Perhaps one of the most common causes of poisoning among sheep is eating the scrapings of paint pots, which have been negligently thrown into corners of pastures, &c. Sometimes the sheep gain access to woodwork, agricultural implements, &c., newly painted, and it is surprising how eagerly they will lick off the application.

Wilful and Malicious Poisoning, happily infrequent, is generally not difficult to trace to the offender. Those who take up this mean calling are usually bunglers, the termination demonstrating forcibly the truth of the old proverb, "Honesty is the best policy."

Symptoms of Poisoning are, as a rule, quickly developed, *generally after a meal;* within a short time after being turned upon a pasture; immediately after the use of a medicinal agent, &c. These facts being kept in mind, we may the more readily direct the attention to the real cause, and determine whether the signs are due to malicious intent, poisonous plants, or mistakes in medicines. Concise information on these points, as far as they are elucidated, should always be sent in writing to the veterinary surgeon when summoned.

The *Treatment of Poisoning* cannot be entered into here; the owner should confine his attempt to alleviate the sufferings of the affected animals. The veterinary surgeon alone can provide the proper antidote. When profuse diarrhœa or dysentery arises, milk with eggs beaten up, thick flour, or starch and water will be useful; and laudanum in 2 to 4 drm. doses may be added if the abdominal pain is acute. If vermin powder has been taken, give broth or soup. When vegetable poisons, as colchicum, hellebore, &c., has been partaken of, add to the milk and eggs nitrous ether, or an ounce of gin, whisky, or brandy. As a vehicle for any of these remedies, one or other of the Demulcents, page 91, may also be used, especially in irritation of the bowels.

CHAPTER XXIII.

Diseases of the respiratory organs—Simple catarrh, or cold—Sore throat—Bronchitis—Inflammation of the lungs—Abscess in the lungs—Pleurisy—Hydrothorax—Pleuro-pneumonia—Asthma—Enzoötic typhoid catarrh, or influenza.

Simple Catarrh,

or *Common Cold*, consists of an inflammation of the lining membrane of the nostrils and sinuses of the head, giving rise to sneezing, or slight cough, a watery discharge, which afterwards becomes purulent, also discharge from the eyes, with more or less sympathetic fever, and even diarrhœa. Catarrh occasionally appears as an enzoötic, depending upon prevailing cold winds, wet, &c., the sheep of whole districts which occupy exposed situations suffering generally. When confined to a few members of a flock, it may be traced to a local cause, as the presence of the larvæ of the *Œstrus Ovis*, or gad-fly. (See chapter xxi.)

Treatment.—Aperients No. 1, page 86, during constipation. If diarrhœa is present at the outset, reduce the oil to one-half, and add 2 drms. of tincture of opium, or the Astringent drench No. 7, page 87, substituting the above dose of oil for water. If the diarrhœa becomes persistent, treat as recommended under that disease, chapter xiv., page 129. Steam the nostrils, and combat fever by Febrifuges, drench No. 2, page 94. When depression ensues, support by nitrous ether in linseed mucilage or flour gruel, as indicated by the state of the bowels. Tonics No. 1 or 2, page 99, when the acute signs have subsided. Good food and shelter are imperative during severe weather.

Sore Throat.

Laryngitis.—Neglected catarrh passes on to sore throat, or it may exist in conjunction from the first. It is frequently an independent affection. The sheep cannot swallow; saliva issues from the mouth when the fingers are inserted to separate the jaws. Pressure on the larynx induces pain, and probably also a painful cough.

Treatment.—Keep the bowels open, by means of Clysters No. 1, page 90; if depression ensues, substitute No. 3. Apply a Blister No. 1, page 87, to each side of the throat, extending from one ear to the other. As the animal cannot swallow, use the Electuary No. 1 or 2, page 92, and, if the breathing becomes very difficult and oppressed, open the windpipe. Tonics may be given when the acute stages are entirely subsided.

Bronchitis.

Inflammation of the Bronchial Tubes.—Owing to prevalence of the same causes which give rise to catarrh and sore throat, bronchitis appears as an enzoötic in sheep from time to time, chiefly during spring and autumn months. There is severe symptomatic fever, distressingly painful cough, inspirations short and difficult, and loud breathing sounds in the bronchiæ.

Treatment.—Bleeding from the jugular when the pulse is full and hard, which is found only at the outset of the malady. Febrifuges No. 2, page 94. Blisters to the sides, front of the chest, and sternum, No. 1, 2, 3, or 4, page 87. If the cough continues, in conjunction with weakness, use Expectorants No. 1 or 2, page 93, twice or thrice a day. When these are overcome, Tonics No. 1, 2, or 3, page 99. Great care is required in this matter, for if iron tonics are used too early the disease may be reproduced in greater severity.

Pneumonia.

Simple Inflammation of the Lungs.—This disease also prevails during severe seasons as an enzoötic, if the situation or district is unusually exposed. It may be the result of aggravation and extension of milder diseases, such as catarrh, &c., particularly bronchitis. It is denoted by intense symptomatic fever, a painful cough, blowing, and more or less loss of breathing space in the lungs, as ascertained by the ear and percussion.

Treatment.—Mild aperients during constipation, No. 1 or 3, page 85. Febrifuges No. 2, page 94. Blisters, as advised for bronchitis, Nos. 1 to 4, page 87, and subsequently as described under that disease.

Abscess in the Lungs.

The short life allotted to the sheep reduces the many affections to which he would be subject if disease were allowed a freer course. On this account abscess of the lungs is seldom seen. It arises in weakly animals after the system is reduced by pneumonia. The disease for a time appears to be of an occult nature, signified by wasting, debility, and weak cough, and finally emaciation and hectic. Further evidences are to be obtained by an examination of the chest by a veterinary surgeon. In later stages an offensive discharge accompanies the cough, and sometimes portions of disorganized lungs are included.

Treatment is tedious and rarely successful.

Pleurisy.

Inflammation of the pleura or lining membrane of the chest sometimes takes place as an extension of disease in pneumonia, &c. It is sometimes independent and enzoötic, and occasionally accompanies rheumatic affections when the heart and its coverings are also implicated. Injuries to the chest also give rise to it. The signs include severe symptomatic fever, abdominal breathing, prolonged expirations, wiry pulse, pain on pressure between the ribs, friction sounds, &c., and negatively, by the absence of disease of the lungs, &c.

Treatment.—Mild Aperients during early constipation, No. 1 or 2, page 86. Strong Febrifuges No. 2, page 94. Blisters to the sides and front of the chest, Nos. 1 to 4, page 87. Blood-letting is of questionable utility. Tonics No. 1 or 2, page 99, as indicated by the reduction of acute disease and restoration of normal frequency in the pulse.

Hydrothorax.

Water in the Chest.—The sudden decline of the active signs of pleurisy are ominous of hydrothorax. The inflammation has been relieved by effusion of water within the cavity of the chest, and this may cause death by pressure on the heart and lungs, thus simulating internal drowning. The presence of water is determined chiefly by estimating and comparing sounds either occurring inside the chest or those elicited by outward percussion.

Treatment.—The operation of tapping is performed for the purpose of drawing off the fluid. This requires the observance of many precautions, for which only the veterinary practitioner is competent.

Pleuro-Pneumonia.

Simple, non-contagious, or sporadic inflammation of the Lungs and Pleura, is known by the arrest of secretions, acute symptomatic fever, full, hard, and rapid pulse, blowing, crepitus or crackling in the lungs during the early stages, friction sounds as in pleurisy, with dead or dull sounds on percussion, and pain between the ribs. There is not the wasting as in contagious pleuro-pneumonia of cattle, and convalescence is frequently established in ten or twelve days.

Treatment as given for pneumonia and pleurisy.

Asthma.

Emphysema of the Lungs.—The use of bulky food containing little nourishment, as an exclusive diet of turnips, &c., necessitates a full stomach, which induces compression of the lungs and interrupts their functions. Enlargement, rupture, and coalescence of the air-cells follow, and the result is asthma. Overfeeding, which produces much fat, induces the same conditions. The lungs lose their elasticity and power; they are inflated with air which occupies ruptured places or false cells. The circulation is impeded, respiration is passive, chest unyielding or fixed, but resonant throughout. There is also weakness and debility, with wasting, arched spine and disinclination to move, pale membranes, &c., in long-standing cases.

Treatment.—There is no cure. Tonics are indicated to promote condition.

Influenza.

See *Enzoötic Typhoid Catarrh*, page 146.

CHAPTER XXIV.

Diseases of the Skin—Simple inflammation, or erythema—Sore teats—Erysipelas—Simple eczema—Chronic eczema—Impetigo larvalis, or black-muzzled ecthyma—Weed—Hidebound—Angleberries, or warts—Foot-rot, or paronychia ovium.

THE above list comprises the non-contagious skin affections. Owing to the thick protective covering, the fleece, the number of diseases of the skin in sheep are materially reduced. Although this animal is highly susceptible to changes of temperature, yet we do not find the same influences exerted through the skin to the same extent as in others having less cutaneous protection and covering.

Erythema,

or *Simple Inflammation of the Skin*, consists of an increase of redness, with moderate swelling in the superficial parts of the skin. It is purely local, and arises from cuts, chafes, moderate blows, and causes of an over-stimulating nature. It is seen between the thighs of males, and between the thighs and udder of females when driven on dusty roads during very hot weather. A chronic form is witnessed in ewes after parturition, due to the effects of wet and cold, &c., when it seizes the udder and forms cracks on the teats.

Treatment.—Both forms give way under cooling lotions, &c. To the first apply Lotion No. 2, page 96; and to the second, No. 7 or 8 Antiseptics, page 85. Aperient No. 2 or 4, page 86, and if the disease assumes chronic forms, use Alteratives Nos. 1 to 4, page 84.

Erysipelas.

This form of inflammation attacks the whole substance of the skin, and occasionally deeper-seated tissues also.

The common causes in sheep are cuts received in shearing; also plethora and sudden check of the usual secretions. There is often much fever, and the inflammation may terminate in abscess.

Treatment.—Febrifuges No. 1 or 2, page 94. Aperients No. 2 or 4, page 86. Lotions No. 2 or 3, page 96, and if necessary, Alteratives Nos. 1 to 4, page 84.

Simple Eczema.

Simple inflammation of the skin (Erythema) sometimes arises in conjunction with the formation of numerous small vesicles or minute bladders. These, by bursting, keep the wool moist, and as an intolerable itching follows their formation, the animal rubs and forms a sore. Newly shorn animals are affected, as a result of the effects of cold. Successive crops of vesicles appear for a time, and thus the disease is continuous.

Treatment.—As for simple erythema.

Chronic Eczema.

Psoriasis, or *Rat Tails.*—This disease is not so well exhibited in sheep as in cattle. When they are subjected to a wet pasture, containing clay, and irritating substances, as lime, &c., particularly in cold weather, the skin of the legs inflames, and the subsequent vesicles, in bursting, influenced by the motions of the limbs, form cracks, which discharge an ichorous fluid. The scaly part of the skin hardens on their edges, and grows rapidly, which makes the cracks appear to be much deeper than they really are. There is much soreness, and even lameness, and, if the malady is neglected, the sheep suffers from the irritation and loses condition.

Treatment.—Fomentations are very useful if continued as recommended at page 94. Next treat as for erythema. Alteratives Nos. 1 to 4, page 84.

Impetigo Larvalis,

Commonly called *Black Muzzle.*—This is a pustular form of eruption, usually confined to lambs while sucking or when folded on long grass. Excoriations and pustules form, and the latter dry up, leaving a dark-coloured encrustation, which, at a later stage, falls off, exposing a red and inflamed surface. It is apt to become chronic and tedious, requiring skill in the treatment. Aperients No. 2 or 4, page 86; reduced for lambs. Astringent lotions or ointments, page 87. Alteratives Nos. 1 to 4, page 84.

Ecthyma.

Another form of pustular inflammation of the skin, confined to those parts which are thin and mobile, as between the thighs and forearms, &c. The vesicles are peculiar, being only few, well defined, isolated, and never confluent. It has been mistaken for small-pox.

Treatment as for the preceding.

Weed.

A dropsical state of the cellular tissue about the udder, and between the hind legs, giving rise to swelling, which, on pressure, retains the imprint of the fingers. It is an accompaniment of other diseases—as mammitis, epizoötic aphtha, &c., and occasionally as a result of debility in the local circulation.

Treatment.—Prolonged fomentations. Internally, Febrifuges No. 1 or 2, page 94. Embrocation No. 2 or 4, page 92, and after the superficial inflammation has subsided apply No. 5.

Hidebound.

This term is used to denote an unhealthy state of the body generally, with deranged functions, such as want of softness and smoothness of the wool, pliancy of the skin, which is bound tightly to the body, as it were. Other organs are always out of order when this condition is present, and its removal entirely depends upon the restoration of the lost or disordered functions, whatever they may be. It must not be estimated nor treated as a distinct disease.

Angleberries.

Or *Warts.*—These are morbid growths, due to causes which generate an inordinate development of one or more parts of the skin. They, therefore, partake of various characters, are non-malignant, but require an operation for their removal; otherwise they crack, become sore, and give much uneasiness, causing loss of condition. When properly extirpated, they do not, as a rule, appear again.

Foot-rot.

Paronychia Ovium.—Inflammation, with abscess of the sensitive structures of the foot. Owing to the insinuation of sand, dirt, &c., combined with the breaking up of the hoof, under the effects of filth and moisture, the discharge becomes black and highly offensive, and the disease chronic.

This malady is said to be contagious under certain circumstances. We have failed to satisfy ourselves as to the truth of this statement, and therefore regard it exactly in the same light as the disease in cattle vulgarly known as "Foot-halt," or "the Low." Its widespread nature, we think, is due to the multiplied causes which soften the hoof and admit of the entrance of irritating substances, leading to inflammation, abscess, and burrowing of matter. This process long continued favours the destruction of bones and ligaments; and a powerful aggravation is no doubt to be found in the septic characters of the surrounding agencies, as manure, urine, &c.

Treatment.—There can be no hope or possibility of

any good arising from whatever measures may be adopted, unless the animals are attended to very early, and they are removed from the pastures where the disease has been contracted. They must be taken to clean lairs, and the floor should be well littered with clean, dry straw. If the sheep could be put on a boarded floor, the joints of which allow the urine to fall through, a still greater advantage would be gained. It is an important item in the process of cure to keep the feet dry after the first operations are concluded.

The dirt and grit contained in the wounds should be removed by fomentations, followed by poultices containing "Sanitas" oil. (See Antiseptics, No. 8, page 85.) Constitutional disturbance must be reduced by febrifuges internally. Remove all loose and obstructing pieces of horn, and open up free passages for the escape of pus when pent up. Amputate diseased portions of bone. When the parts are clean, and free from irritating substances as the foregoing, promote the healing process as quickly as possible. Sinuses syringed with caustic solution No. 6, page 89; and as a general application the Healing Fluid No. 4, page 96, will be found efficacious. Dressings of "Sanitas" oil may be applied by means of tow secured by narrow bandages passed round the foot and between the hoofs.

CHAPTER XXV.

Diseases of the urinary organs—Diabetes—Retention of urine—Incontinence of urine—Simple albuminuria—Hæmaturia, or bloody urine—Sthenic Hæmaturia—Inflammation of the kidneys—Inflammation of the bladder—Gravel and calculi, or stone—Protrusion and inversion of the bladder—Weakness of the bladder—Discharge of pus from the bladder.

In the numerous diseases in which the urinary organs are implicated the causes are originally due to errors of diet, the remainder being produced by violence. The kidneys are closely connected by their functions with the processes of digestion and assimilation; many substances incapable of being appropriated by those processes are transferred to the kidneys for elimination. It will, therefore, be obvious that when unnatural elements of food are constantly supplied to the stomach, the functions of the kidneys and allied organs will be proportionately deranged, and by long continuance of disorder eventually diseased.

Diabetes.

Profuse Urination, in conjunction with general debility and loss of condition arising from improper food, drugging, and defective management generally. It also appears as a sequel of long-standing and wasting diseases.

Treatment.—Mild aperients, Nos. 1 to 3, page 86; subsequently iodine, phosphate of iron, &c., under veterinary advice.

Retention of Urine.

An accumulation and retention of urine in the bladder arises from spasm or inflammation of the neck, paralysis of the bladder itself, stone, inversion of the rectum or vagina, and from pressure of other organs. Male animals are more frequently affected.

Treatment.—Pass the catheter speedily, or the bladder

may be ruptured, causing death. In the male an operation must be performed if the cause of retention is with the bladder or its neck.

Incontinence of Urine.

Dropping or dribbling of the urine from the penis or vulva of adult animals, and in lambs from the navel. It is due to paralysis or debility in the former, and to imperfect closure of a fœtal canal—the urachus in the latter.

Treatment.—Tonics No. 1 or 2, page 99, or strychnine for adult sheep. For lambs, put a ligature on the navel. Tonics. See "Weakness of Bladder."

Simple Albuminuria.

Albuminous Urine, arising from inflammation of the substance of the kidney, giving rise to pain, straining, arched back, and feet drawn together. Epithelial scales and granular matter are also found in the urine.

Treatment.—Aperients Nos. 1 to 3, page 86, and Clysters No. 1 or 2, to preserve a uniform state of the bowels. Mustard to the loins. Astringents No. 7 or 8.

Hæmaturia.

Or *Bloody Urine*. As a result of injury causing structural changes in the kidneys, blood is separated with the urine. The discharge is accompanied with much pain and sympathetic fever, difficulty in walking, and even paralysis.

Treatment.—Cold-water clysters; wet rags over the loins, or a stream of cold water. Aperients are not always needed or even safe, except No. 1, page 86. Pass the catheter if the bladder is full. Astringents internally, No. 7 or 8, page 87.

Sthenic Hæmaturia.

An enzoötic affection, usually chronic in its nature, due to the action of acrid poisonous plants taken as food.

The signs resemble those given under the last form, but are slowly developed, and the colour of the urine is orange-red, often transparent unless large quantities of carbonates are present.

Treatment. — Aperients No. 4, page 86. Clysters No. 1 or 2, page 90. Mustard to the loins. Linseed mucilage as a drink. Astringents No. 7 or 8, page 87.

Inflammation of the Kidneys.

Nephritis.—The signs are abdominal pain, arched back and straining, intense symptomatic fever, albuminous urine without blood at first, but in later stages it is voided with pus in addition. Diarrhœa, dysentery, or uræmia usually terminate the disease.

Treatment.—Aperients No. 1 or 2, page 86, with belladonna. Warm Clysters No. 1, page 90. Febrifuges No. 2, page 94. Mucilaginous drinks. Mustard to the loins. Astringents No. 7 or 8, page 87.

Inflammation of the Bladder.

Cystitis.—Great pain and uneasiness, with efforts to vomit, attend this disease. The urine is loaded with mucus, shreds of fibrine, and epithelial debris. Sometimes the flow is unimpeded, at others entirely suspended, owing to spasm at the neck. In the latter instance the bladder will be full and in danger of rupture.

Treatment.—Evacuate the bladder as speedily as possible if it is full. Aperient No. 3 or 4, page 86. Warm Clysters No. 1, page 90. Demulcents Nos. 1 to 4 should be used as vehicles for the medicines, and No. 1 or 3 injected into the bladder if no spasm exists. Mustard embrocations to the loins. Febrifuges No. 3, page 93. If stone is present, its removal must be decided.

Gravel.

Accumulations of sandy matter occasionally take place in the bladder, and give rise to much uneasiness. Male animals are mostly sufferers. Stones, or calculi, also form, producing inflammation, and interfering with the flow of urine. Smaller stones lodge in the urethral canal, or with gravel block up the orifice at the prepuce. For each of these the assistance of the veterinary surgeon is needed. Limestone districts furnish many cases of this kind. The land requires potash dressings; drinking water should be neutralised by soluble carbonates, and root and other crops from fertile soils should be supplied.

Protrusion and Inversion of the Bladder.

This is a serious accident under all circumstances. It occurs only in adult ewes or ewe lambs. In one form it is literally turned inside out, and in the second it is forced through a rupture in the floor of the vagina, constituting a variety of hernia. The organ must be returned to its original situation, and the breach in the vagina closed by sutures. For these the veterinary surgeon should be consulted.

Inversion of the bladder may arise in consequence of severe straining during parturition, and as a result of the use of powerful diruetic medicines given to adults. The unauthorised practice of giving medicines by cowmen, shepherds, and others, who know literally nothing of the nature of the substances, require to be met by stern treatment. It should be a standing rule in every farm and establishment where animals are kept, that whoever administers drugs of any kind to any of those animals—whether horse, cow, dog, or sheep—without due permission, or on the authority of certain regulations, the result should be dismissal. From this propensity for drugging many cases are seriously complicated, and others terminate fatally. With little trouble and cost, owners of stock, living far away from towns and veterinary assistance, might be provided with safe compounds for all ordinary emergencies. The veterinary surgeon in supplying these would furnish ample instructions, so that under the owner's superintendence there might at least be reason as well as safety in the use of medicines. The question which should be seriously considered, is the great desirability of retaining in every country district a veterinary practitioner, whose whole time might be devoted to a certain number of farms, with the view of practising *prevention* of disease. Thus the salary of an efficient man would be saved, doubled, and even quadrupled by the farmers, and the nation all the better for the supplies. See chapter xxviii. of "Cattle, their Varieties in Health and Disease," uniform with this work.

Weakness of Bladder.

Sheep in low condition, having passed through a trying winter or a prostrating disease, are sometimes unable to discharge the contents of the bladder in a continuous stream. There is no evidence of pain, but the attempts to urinate are frequent—the animal stands a long time passively prepared, without straining, and succeeds only in discharging a small quantity in an intermittent flow. The cause is a want of tone in the bladder, notably the muscular coat, arising from the general debility of the system.

Treatment.—Create a regular condition of the bowels by frequent use of Clysters No. 1, page 90, and give linseed mucilage freely as a drink, with green food, roots, &c., in addition to other nourishing articles of diet. The medicines should comprise Tonics No. 1 or 2, to which a small dose of cantharides may be added in the form of fine powder, until the proper action of the bladder is restored. We must remind the reader of the uncertain and dangerous effects of this remedy if given irregularly and incautiously, or continued too long. The results may be irritation and inflammation of the kidneys or bladder, which will frustrate the end which is sought after. When weakness of the bladder follows previous inflammation of any of the urinary organs, cantharides must be used only under great watchfulness, or altogether abandoned. A full history of the case should be given to the veterinary surgeon, who would then attempt other means, and avoid reproducing the original malady.

Discharge of Pus from the Bladder.

If signs of fever, consequent upon local inflammation, are present, the cause should be first removed by means of aperients, if needful (p. 86), and sedatives (No. 2, p. 94). As a simple complaint, and dissociated from febrile action, the discharge may be arrested by two or three medium daily doses of spirits of turpentine, each beaten up with a couple of eggs, and given in linseed mucilage. The oil of eucalyptus, or "Sanitas" oil (ten to twenty drops) may be substituted for the turpentine, especially in chronic cases, and alternated with iodide of iron in combination with cantharides and gentian. Local remedies should also be injected into the bladder of the ewe, as eucalyptus oil, "Sanitas" oil, or weak solutions of the sulphate or chloride of zinc. In the ram this is difficult and sometimes impossible without a special operation, for which the veterinary surgeon will be needed. Allow demulcent drinks (p. 90) ad libitum, and administer the medicine in the same, instead of water.

INDEX.

ABDOMEN, dropsy of, 159
„ injuries to, 160
Abortion, 17, 153
Abscess, 75
„ in the lungs, 198
Accidental poisoning, 194
Acute dysentery, 129
„ indigestion, 125
After pains, 138
Argali, the, 1
Agricultural wealth, sheep a source of, 3
Albuminuria, 207
Alteratives, 84
Anæmia, 101
Anæmic palpitation, 108
Angleberries, 204
Anodynes, 84
Anthracoid diseases, 138
Antiputrescents, 85
Antiseptics, 85
Antispasmodics, 86
Aperients, 86
Aphtha, epizoötic, 113
Apnœa, 105
Apoplexy, 167
„ before parturition, 138
„ pre-parturient, 138
„ splenic, 136
Arsenical wash, 193
Arteries and veins, wounds of, 162
Arthritis, 144
Asthenic hæmaturia, 145
Asthma, 200
Astringents, 86
August, the flock in, 53
„ weaning lambs in, 53
Autumn, the sheepfold in, 58
Australian experiences of sheep washes, 46
BELGIAN carrot as food, 61
Big-headed sheep, 3
Big-tailed sheep, 2
Black v. Elliott, alleged sheep poisoning case, 47
Blackleg, 134
Black muzzle, 203
„ quarter, 134
„ spauld, 134
„ water, 145
Bladder, discharge of pus from, 211
„ inflammation of, 208
„ inversion of, 209
„ protrusion of, 209
„ sand in, 209
„ weakness of, 210
Blain, 137
Blasted, 125
Bleeding cancer, 152
„ from the vagina, 156
„ „ womb, 156
Blisters, 87
Blood disease in lambs, 149
„ diseases, 99, 113, 142
„ fungus, 152
„ in the urine, 207
„ poisoning, 103
„ striking, 134
Blown, 125
Blue disease, 109
Blundell, Mr., on the Belgian carrot as food, 61
Bone, fractures of, 165
„ softening of, 107
Bowels, wounds of, 162
Brain, hydatid disease of, 174
„ inflammation of, 167
„ water on the, 168
Braxy, 137
Breaking-down, 164

Breathing, difficult, 105
Breeding ewes, feeding of, 14
,, flock, 14
Bronchial tubes, inflammation of, 197
Bronchitis, 197
,, verminal, 173
Brown, Professor, on the fluke disease, 178
Bulging of the cornea, 152

CANCER of the orbit, 152
Carbuncular erysipelas, 134
Carditis, 109
Castration, 19
Catarrh, enzoötic typhoid, 146
,, malignant, 143
,, simple, 196
Causes of abortion, 154
Caustics, 89
Characteristics of sheep, 2
Chine felon, 102
Chest, water in, 199
Choking, 126
Chronic dysentery, 130
,, eczema, 202
,, hoven, 125
,, indigestion, 128
Circulatory system, diseases of, 108
Cleansing, 155
Clysters, 90
Cobbold, Dr., on tapeworm, 175
Cold, common, 196
,, felon, 102
Colic, 129
Common cold, 196
Composition of the wash, 43
Consumption, tubercular, 184
Contagion, prevention of, 121
Contagious diseases, 113
Contused wounds, 164
Cordials, 90
Cornea, bulging of, 152
Cost of sheep shearing in Gloucester, 52
Cost of sheep shearing in Sussex, 52
Cotswold sheep, 7
Cud, dropping of the, 128
Cyanosis, 109
Cyst, serous, 76
Cysticercus tennuicolis, 176
Cystitis, 208

DECEMBER, the flock in, 61
Demulcents, 90
Dentition of the sheep, 12
Derbyshire neck, 106
Dew blown, 125
Diabetes, 206
Diaphoretics, 91
Diarrhœa, 129
Deficiency of blood, 101
Difficulty in breathing, 105
Digestive canal, worms in, 190
,, organs, diseases of, 124
Digestives, 91
Dilatation of the heart, 112
Dipping, 192
Discharge of pus from the bladder, 211
Disease, conditions of, 71
,, the fluke, 177
,, prevention of, 69
Diseases, anthracoid, 138
,, anthrax, 134
,, of the blood, 99, 113, 142
,, ,, circulatory system, 108
,, contagious, 113
,, digestive organs, 124
,, enzoötic, 133
,, epizoötic, 113
,, of the eyes, 150
,, ,, generative organs, 153
,, ,, nervous system, 167
,, parasitic, 172
,, of the respiratory organs, 196
,, ,, skin, 201
,, ,, urinary organs, 206
Disinfection, 122
Dislocations, 166
Displacement of the heart, 112
Disposal of manure, 122
Distoma Hepaticum, 178
Diuretics, 91

Domestic animals, value of, 65
,, sheep, 2
Drafting the feeding flock, 18
Dressing for foot-rot, 55
Dropping of the cud, 128
Dropsy of the abdomen, 151
,, sanguineous abdominal, 148
,, of the udder, 203
Dun's sheep wash, 43
Dutch clover for lambs, 40
Dysentery, acute, 129
,, chronic, 130

EARLY lambs, 14
Echinococcus parasitism, 176
Ecthyma, 203
Eczema, simple, 202
,, chronic, 202
Electuaries, 92
Embolism, 113
Embrocations, 92
Emphysema of the lungs, 200
Empirical poisoning, 193
Endocarditis, 111
Enemas, 90
Enlargement of the heart, 112
Enteritis, 130
Enzoötic diseases, 133, 142
,, typhoid catarrh, 146
Ephemeral fever, 72
Epilepsy, 168
Epistles, laconic, 82
Epizoötic aphtha, 113
,, diseases, 113
Erysipelas, 201
,, carbuncular, 134
Erythema, 201
Ewe flock, practical farmer on, 34
,, ,, the, 47
Ewes, Belgian carrot for, 61
,, second season, 48
Expectorants, 93
Eye, diseases of, 150

FALSE presentations, 17
,, positions of the lamb, 156
Fardel-bound, 127
Fatting of lambs, 38
,, tegs, 57

Fatty degeneration of the heart, 112
Febrifuges, 94
February lambing, 23
Feeding flock, drafting of the, 18
,, of lambing ewes, 14
,, in stalls, 32
,, sheep in yards, 27
Fever, 71
,, ephemeral, 72
,, inflammatory, 134
,, parturient, 139
,, puerperal, 17
,, simple, 72
,, specific, 74
,, sympathetic, 72
,, symptomatic, 72
Finlay Dun's sheep-wash, 43
Five suggestions, 87
Flemish sheep, 9
Flock, the breeding, 14
,, in January, 14
,, ,, February, 23
,, ,, March, 34
,, ,, April, 37
,, ,, May, 38
,, ,, June, 41, 47
,, ,, July, 52
,, ,, August, 53
,, ,, September, 54
,, ,, October, 58
,, ,, November, 60
,, ,, December, 61
Flooding, 156
Fluke disease, 177
,, ,, Professor Brown on, 178
,, ,, Professor Simonds on, 178
Fly, the, 42
,, mixture, 46
Food and increase, 63
Fog sickness, 125
Fomentations, 94
Foot-halt, 204
Foot and mouth disease, 113
Foot-rot, 55, 204
Foreign bodies in the heart, 110
,, ,, ,, rumen, 126
Fractures of bone, 165

Fulness of blood, 99
Fungus Hæmatodes, 152

GAD-FLY, attacks of, 172
Garget, 159
General wounds, 163
Generative organs, diseases of, 153
Gid, the, 174
Glos Anthrax, 137
Gloucester, sheep shearing in, 52
Godham's salve, 45
Goggles, 174
Goitre, 106
Grain-sick, 126
Gravel, 209
Grub in nasal sinuses, 172

HACKS, removal of, 153
Hæmaturia, 207
,, asthenic, 145
,, sthenic, 207
Hæmorrhage, vaginal, 156
Hampshire Farmer on feeding ewes and lambs, 22
Hasty, 134
Haw, removal of, 153
Healing lotion, 96
Heart-bag, inflammation of, 110
,, dilatation of, 112
,, displacement of, 112
,, enlargement of, 112
,, fatty degeneration of, 112
,, foreign bodies in, 110
,, inflammation of, 109
,, rupture of, 109
Heaving pains, 139
Hidebound, 203
Hogg's ointment, 44
,, salve, 45
Honey-comb bag, diseases of, 127
Hoose or husk, 56, 173
Hoove or hove, 125
Hoven, chronic, 125
Husk, 173
Hydatids in the brain, 174
Hydrocephalus, 168
Hydrocephalus hydatidæus, 174
Hydro-rachitis, 169
Hydrothorax, 199

ICELAND sheep, 3
Impaction of the rumen, 126
Imperfect messages, 78
Impetigo larvalis, 203
Incised wounds, 163
Incontinence of urine, 207
Increase from food, 63
Indigestion, acute, 125
,, chronic, 128
Inflammation, 74
Inflammation of the bladder, 208
,, ,, brain, 167
,, ,, bronchial tubes, 197
,, ,, heart, 109
,, ,, heart-bag, 110
,, ,, intestines, 130
,, ,, kidneys, 208
,, ,, lungs, 198
,, ,, ,, and pleura, 199
,, ,, liver, 132
,, ,, peritoneum, 131
,, ,, pleura, 198
,, ,, skin, 201
,, ,, udder, 159
,, ,, urethra, 158
,, ,, vagina, 158
Inflammatory fever, 134
Influenza, 146
Injuries, local, 16
,, to the abdomen, 160
,, ,, mouth, 162
Intestines, inflammation of, 130
Inversion of the bladder, 209
,, ,, vagina, 157
,, ,, womb, 157
Iritis, 151

JANUARY lambing, 14
Jaundice, 132
Joint disease in lambs, 144
,, felon, 102
,, ill, 134
Joints, punctures of the, 164
June, sheep dipping in, 52
,, the flock in, 45

KIDNEYS, inflammation of, 208

LABOUR, natural, 155
Lacerated wounds, 163
Laconic epistles, 82
Lamb, false positions of, 156
Lambing, 16, 35
Lambing in January, 14
,, ,, February, 23
,, on lowland farms, 23
,, oils, 141
,, in south of England, 53
Lambs, blood disease in, 149
,, castration of, 19
,, early, 14
,, fatting, 38
,, joint disease in, 144
,, peas for, 38
,, weaning, 51
Lammermoor farm notes, 36
,, management, 23
,, operations in September, 57
,, operations in October, 59
,, report in November, 63
,, sheep farms, 24
Lammermoors, weaning lambs on, 54
Laryngitis, 197
Leaping ill, 109
Leicester sheep, 6
Lice, 193
Liniments, 92
Liver fluke, the, 178
,, inflammation of, 132
,, parasitic disease of, 176
Local injuries, 160
Locked jaw, 170
Lotions, 95
Louping ill, 169
Low, the, 204
Lowlands farm, lambing on, 23
Lungs, abscess in, 198
,, empysema of, 200
,, consumption of, 104
,, inflammation of, 199
,, simple inflammation of, 198

MAGGOTS, 193
,, a remedy for, 46
Malicious poisoning, 194
Malignant catarrh, 143
,, sore throat, 145
Mammitis, 159
Mange, 191
Manure, disposal of, 122
MacLagan's flock, 15
March lambing, 34
Martin's sheep-wash, 44
Materia Medica, 69
Mawbound, 126
Measles, 120, 175
Medicated poultices, 93
Membranes, expulsion of, 155
Merino sheep, 10
Messages imperfect, 78
Mission of veterinary science, 69
Morbid pathology, 69
Mouth, injuries to, 162

NASAL sinuses, grub in, 172
Natural labour, 155
Nature and qualities of wool, 4
Navel ill, 149
Nephritis, 208
Nervous system, diseases of, 167
Notes from Lammermoor farm, 36
November report from Lammermoor, 63
,, sheepfold in, 60

OCTOBER, operations in, 59
Œstrus Ovis, 172
Oil-cake, 39
,, and peas, for ewes and lambs, 22
Oils, lambing, 141
Operations on Lammermoor in September, 57
Operations on Lammermoor in October, 59
Ophthalmia, simple, 150
Orbit, bleeding cancer of, 152
Organs of circulation, diseases of, 108
,, ,, digestion, diseases of, 124

Organs of generation, diseases of, 153
„ „ the nervous system, diseases of, 167
„ „ respiration, diseases of, 196
„ „ the skin, diseases of, 201
„ „ urination, 206
„ „ vision, 150
Origin of the sheep, 1
Outbreaks of fluke disease, 177
Ovis Ammon, 1

PALPITATION, 108
Palsy, 169
Paralysis, 169
Parasitic disease, 172
„ disease of the liver, 176
Parasitism, echinococcus, 176
Parturition, apoplexy before, 138
„ fever, 139
Pathology, 69, 71
Peas for lambs, 38
„ white, *v.* maple, 39
Percentage of increase in fatting sheep, 64
Pericarditis, 110
Peritonitis, 13?
Phrenitis, 167
Phthisis, 104
Plethora, 99
Pleura, inflammation of, 198
Pleurisy, 198
Pleuro-pneumonia, 199
Pneumonia, 198
Poisoning, accidental, 194
„ empirical, 193
„ malicious, 194
„ symptoms of, 195
„ treatment of, 195
„ wilful, 194
Poisons, 193
Poultices, 96
Pouring, 192
„ for ticks, 59
Practical Farmer on the ewe flock in March, 34
Pre-parturient apoplexy, 138
Prevention of contagion, 121
Prevention of disease, 69
„ „ the fly, 42
Protrusion of the bladder, 209
Profuse urination, 206
Psoriasis, 202
Public health in relation to the health of stock, 67
Puck, 134
Puerperal fever, 17
Punctured joints, 164
„ wounds, 164

QUARTER evil or ill, 134

RABIES, 170
Rams, shearing, 49
Rat tails, 202
Red water, 145, 148
Respiratory organs, diseases of, 196
Reticulum, diseases of, 127
Retention of urine, 206
Retinitis, 151
Rheumatism, 102
Rickets, 107
River-fluke, 178
Robertson, W., on plethora, 100
Root feeding of ewes and lambs, 20
Rot in sheep, 177
Rubeola, 120
Rumen, foreign bodies in, 126
„ impaction of, 126
Rupture of the heart, 109
„ „ womb, 157
Ruston, Mr., on feeding sheep in yards, 27

SALVE, the, 42
Salving, 45, 192
"Sanitas," a disinfectant, 123
Sanguineous abdominal dropsy, 148
Sauntering, 153
Scab or mange, the, 42, 191
„ U.S., remedy for, 46
Second season ewes, 48
„ stomach, diseases of, 127
Sending for the veterinary surgeon, 77

September, the flock in, 54
,, operations on Lammermoor, 57
Serous cyst, 76
Shearing rams, 49
Shed-feeding, 32
Sheep, a source of agricultural wealth, 39
,, big-headed, 3
,, big-tailed, 2
,, characteristics of, 2
,, dentition of the, 12
,, dipping, 192
,, ,, in June, 52
,, domestic, 2
,, origin of, 1
,, fairs, 10
,, farm, Lammermoor, 24
,, Iceland, 3
,, ointment, Hogg's, 43
,, ,, Youatt's, 44
,, pouring for ticks, 59, 192
,, rot, 177
,, salving, 192
,, shearing, 41, 52
,, smearing, 192
,, various breeds of, 5
,, Wallachian, 3
,, wash, 42
,, ,, Finlay Dun's, 43
,, ,, Martin's, 44
,, ,, Southdown Company's, 46
,, washes, Australian experience of, 46
,, washing, 41
Shelter for the shepherd, 16
Sheepfold in autumn, 58
,, November, 60
Shewt or shoot of blood, 134
Simonds, Professor, on the fluke disease, 178
Simple albuminuria, 207
,, catarrh, 196
,, eczema, 202
,, fever, 72
,, inflammation of the lungs, 198
,, inflammation of the skin, 201
Simple ophthalmia, 150
Skin, simple inflammation of, 201
Slinking, 153
Slipping the lamb, 153
Shropshire sheep, 9
Small-pox, 117
Smear, the, 42
Smearing, 192
Softening of bone, 107
Sore throat, 197
,, ,, malignant, 145
Southdown Company's sheep-wash, 46
,, sheep, 8
South of England, lambing in the, 53
Specific fever, 74
Speed, 134
Splenic apoplexy, 136
Spongio piline, 90
Spooner, V. S., his recipe for foot-rot, 55
,, on the hoose, 56
Sporadic aphtha, 124
,, inflammation of the lungs, 199
Sprain of tendon, 165
Stall feeding, 32
Staphyloma, 152
Sthenic hæmaturia, 207
Stomach staggers, 127
Sturdy, 174
Summer, the flock in, 51
Sussex, sheep-shearing in, 52
Sympathetic fever, 72
Symptomatic fever, 72
Symptoms of poisoning, 195

Tænia echinococcus, 176
,, *marginata*, 176
,, *tenella*, 174
Tapeworm, Dr. Cobbold on the, 175
Tegs, fatting, 57
Tendon, sprain of, 165
Tetanus, 170
Third stomach, impaction of, 127
Throat, sore, 197
Throttle ill, 170
Thrush of the mouth, 124

Thumps, the, 108
Thwartil ill, 170
Ticks, 193
„ pouring for, 59
Tonics, 98
Treatment of lambing ewes, 35
„ „ poisoning, 195
Trembling, 169
Tubercular consumption, 104
Tuberculosis, 104
Turnsick, 174
Turnside, 174
Tuson's arsenical wash, 193

UDDER, inflammation of, 159
United States remedy for scab, 46
Unnatural position of the lamb, 156
Uræmia, 103
Urethritis, 158
Urination, profuse, 206
Urine, albuminous, 207
„ blood in, 207
„ incontinence of, 207
„ retention of, 206
Urinary organs, diseases of, 206
Utility of wool, 4

VAGINA, inversion of, 157
„ inflammation of, 158
Vaginal hæmorrhage, 156
Vaginitis, 158
Value of domestic animals, 65
Various breeds of sheep, 5
Veins, wounds of, 162
Ventilation, 121
Verminal bronchitis, 173
Vertigo, 127, 174
Veterinary medicine, 70
„ science, mission of, 69
„ surgeon, sending for, 77
„ surgery, 70

WALLACHIAN sheep, 3
Warping, 153
Warts, 204
Water on the brain, 168
„ in the chest, 199
Weakness of the bladder, 210
Weaning lambs, 52
„ on the Lammermoors, 54
Weed, 203
Welsh sheep, 9
Wilful poisoning, 194
Winter purchases of stock, 54
Womb, bleeding from, 156
„ inversion of, 157
„ rupture of, 157
Wool, nature and qualities of, 4
„ utility of, 4
Worms in the digestive canal, 190
Wounds, 163
„ of arteries and veins, 162
„ „ the bowels, 162

YARDS, feeding sheep in, 27
Yellows, the, 132
Youatt's ointment, 44
Young, Mr., his experience of sheep-washes in Australia, 46

PRINTED BY BALLANTYNE, HANSON AND CO.
LONDON AND EDINBURGH

THE HORSE

ITS VARIETIES AND MANAGEMENT IN HEALTH AND DISEASE

With Numerous Page Plates and other Illustrations.

THE colossal proportions attained by the horse stock of the United Kingdom, together with its almost fabulous money value, are incentives to an inquiry into the maladies common to the race. But without these humanity is bound to the task in return for a service, the intelligence of which is not equalled by other domestic animals.

The maladies of the horse are numerous and largely fatal, yet under a possible system of management they are mainly preventable. The object of this treatise is to render these truths clear and acceptable, proving itself the adviser of the farmer and horse owner at home or abroad.

With this view the part on horse management is treated in a full and comprehensive manner.

The list of indigenous diseases has been compiled with care; and that dealing with contagious maladies is also full and complete, with ample instructions for medical and other treatment in every instance.

Under a special arrangement the various remedies are given in conjunction with precise directions as to dose, form of combination, &c., readily understood by all readers. Treatment by subcutaneous injections forms the subject of a separate section.

Local injuries and lameness are noticed as far as space will admit, and the work concludes with a special chapter on shoeing, in which the best system of management of the feet is embodied.

CATTLE

THEIR VARIETIES AND MANAGEMENT IN HEALTH AND DISEASE.

With Numerous Page Plates and other Illustrations.

THE maladies of horned stock are numerous and important. Their tendency to fatal terminations has largely diminished the productive capabilities of the country, British wealth being transferred to the coffers of the foreign breeder and importer. It is also understood that the well-being of cattle and sheep greatly determines the safety of the public health. With these premises in view, an increasing intelligence welcomes the means of investigating, and is already accepting the proposition that the application of measures, the outcome of sound conclusions, have the certain effect of mitigating and even suppressing the evils referred to.

The present work is intended as an adviser of the farmer and breeder in that particular direction.

On the subject of disease the information is copious, comprising not only the more common, as indigenous to our soil, but also those which reach us through the system of free intercourse with pest-ridden countries abroad. These are grouped under an arrangement in conformity with their acknowledged nature, location, and causes. The symptoms are carefully noted throughout ; full instructions as to treatment are given, and a special section is devoted to the suitable remedies.

Suggestions for a complete medical supervision are advanced, with the object of leading up to an efficient control over the prevalence of disease, but primarily to prevent its origin. This is a department of untold benefit to our country, and is capable of great development.

SHEEP

THEIR VARIETIES AND MANAGEMENT IN HEALTH AND DISEASE

With Numerous Page Plates and other Illustrations.

THE diseases of sheep, in common with those of cattle, exhibit special peculiarities with respect to nature, prevalence, and fatality. Many are preventable, having their origin in the character of soil, or conditions of general management. During recent years greater attention has been paid to the subject, with the result that important light has been thrown on its various aspects.

The development of a gigantic industry in New Zealand, Australia, the Argentine Republic, and Cape Colony, with the view of furnishing food for the millions at home, calls for persistent investigation among the sheep raised under specific conditions of climate, soil, vegetation, &c., of these countries. Vast areas of land, hitherto undisturbed, are now occupied by numberless flocks imported from distant homes. The greatest diversity thus exists, and the results are but faintly appreciated, or altogether uninterpreted. They finally operate adversely to successful breeding, and sometimes render the industry abortive. But the loss is not confined to the breeder and sheep farmer. The health of the people is in direct ratio to the soundness of the flesh offered for their daily food. Successful sheep raising, whether at home or in the colonies and elsewhere, is a boon and a blessing to the nations in proportion as the farmer is blessed, and its maintenance as such alone depends upon the intelligence with which the subject is investigated and pursued.

These are cogent reasons for an inquiry into all that militates against the standard of health, and the present work purports to be a guide to the sheep farmer in that especial direction.

In the classification of diseases, their nature, location, and mode of origin have been taken as the basis—a plan which will be found simple and effective.

The symptoms are compiled with care, and the details of medical treatment are ample and precise.

A complete section is devoted to remedies and the various forms of administration.

The contagious diseases of the sheep, and those also due to parasitism, are carefully described—to which, appropriate measures for alleviation and prevention are appended. In all respects the work is abreast of present-day experience, and is a useful adviser in the emergencies common among the stock of the fold.

These Works can be obtained at all Booksellers in the United Kingdom and Colonies,

OR DIRECT FROM THE PUBLISHERS,

FREDERICK WARNE AND CO.

BEDFORD ST., STRAND, LONDON; & NEW YORK.

www.ingramcontent.com/pod-product-compliance
Lightning Source LLC
Chambersburg PA
CBHW021519210326
41599CB00012B/1310
* 9 7 8 3 7 4 2 8 4 2 5 4 1 *